T0269234

Finding the Nerve

Finding the Nerve
The Story of Impedance Neurography

Philip C. Cory
President & Patient Partner
Beargrass Patient Partners, PLLC

Academic Press is an imprint of Elsevier
125 London Wall, London EC2Y 5AS, United Kingdom
525 B Street, Suite 1800, San Diego, CA 92101-4495, United States
50 Hampshire Street, 5th Floor, Cambridge, MA 02139, United States
The Boulevard, Langford Lane, Kidlington, Oxford OX5 1GB, United Kingdom

Notices
Knowledge and best practice in this field are constantly changing. As new research and
experience broaden our understanding, changes in research methods, professional practices,
or medical treatment may become necessary.

Practitioners and researchers must always rely on their own experience and knowledge in
evaluating and using any information, methods, compounds, or experiments described
herein. In using such information or methods they should be mindful of their own safety and
the safety of others, including parties for whom they have a professional responsibility.

To the fullest extent of the law, neither the Publisher nor the authors, contributors, or editors,
assume any liability for any injury and/or damage to persons or property as a matter of
products liability, negligence or otherwise, or from any use or operation of any methods,
products, instructions, or ideas contained in the material herein.

Library of Congress Cataloging-in-Publication Data
A catalog record for this book is available from the Library of Congress

British Library Cataloguing-in-Publication Data
A catalogue record for this book is available from the British Library

ISBN: 978-0-12-814176-2

For information on all Academic Press publications visit our
website at https://www.elsevier.com/books-and-journals

  Working together
to grow libraries in
Book Aid
International developing countries

www.elsevier.com • www.bookaid.org

Publisher: Nikki Levy
Acquisition Editor: Natalie Farra
Editorial Project Manager: Pat Gonzalez
Production Project Manager: Kiruthika Govindaraju
Designer: Matthew Limbert

Typeset by TNQ Books and Journals

*To the memory of Patrick Wall, DM, FRS
who, when presented with initial data,
was the first to recognize the potential
of Impedance Neurography and to Joan
Cory, PhD, for her inventive contributions,
insights, counsel, and unflagging support.*

Contents

4. Anisotropicity

5. Depth Determination of Peripheral Nerves Using
    Impedance Neurography

Preface

In this work, I describe a technology that grew out of an urge to find better ways to investigate chronically painful conditions and provide people suffering from them with improved symptomatic relief or cures. It has been a decades long scientific investigation that involved learning many lessons about neuroscience and business, while meeting very well-qualified and capable people who helped with all aspects of the project.

Though we have not been able to commercialize this technology, it is my intent that the information be available to the research and development community, so they may find it possible to pursue further work in this area. I hope this small volume serves to pique interest in the technology, its potential for helping people, and provides interesting stories during the reading of some of the drier bits. I also hope that the insights developed in researching the underlying nature of the electroneurophysiology can assist those working in related areas such as Impedance Tomography, which has yet to fulfill its theoretical promise. Understanding the electrical role that nerves play in living tissue is key to recognizing the way electrical fields distribute in living bodies and how it differs from cadaveric tissue. This work contains insights regarding aspects of electrotonic nerve conduction with applications to neuromodulation technologies, and perhaps even to the goal of electroanesthesia. There is also much to be developed in the mathematical description of the effect, and it is my hope that researchers will pick up where I have left off to complete the picture.

Philip C. Cory, MD

Acknowledgments

This work would not have been possible without the help the many people who sought my advice in attempting to resolve their chronic pain problems. Their willingness to allow me to use prototype Impedance Neurography devices to image nerves and nerve-related abnormalities was invaluable to discovering the nuances of developing a useful medical instrument and technology.

I am also very grateful to Dr. John Miller, PhD, Dr. James McMillan, PhD, and Dr. Murari Kajariwal, PhD for their advice, instruction, and patient review of ideas and written work. Dr. Dean Shultz, PhD provided the computer programming experience necessary for developing an excellent proof of concept device in concert with the very capable work of the development team under the able direction of Dr. Daniel Kramer, PhD at Battelle Memorial Laboratories in Columbus, Ohio. The investigations could never have gotten off the ground without financial support in the form of a grant from B. Braun, Inc., facilitated by Mr. Brad Lane as well as the financial support of the investors in Nervonix, Inc. and grants from the Montana State Department of Commerce. Finally, I am most grateful to the late David Simons, M.D. who flew so high in his career, began as a nonbeliever but came to be an enthusiastic advocate of Impedance Neurography. Over the years spent in the development of a working hypothesis describing the underlying neurophysiology enabling Impedance Neurography these people have provided essential support, without which this book would not have been possible.

Introduction

In many areas of medicine and biomedical research, the ability to image nerves in the living person or animal has been a long sought-after goal. This capability would be helpful for finding nerves as a target tissue, e.g., for regional anesthesia or neurodiagnostic studies, or for actively avoiding nerves during procedures that have the potential to damage them. Impedance Neurography does exactly that; it provides an image of peripheral nerves in the living state, showing both those nerves that appear to be normal and those demonstrating abnormal anatomy or function. This is fascinating stuff, and years ago, when we presented our preliminary findings to a possible partnering company the comment was made that if we could actually do this, we had the "holy grail" of neurology. It turned out that we could do it, and the process of learning how we did it opened up a host of new applications as well as offering insights as to how to improve existing technologies.

This book details the development of an innovative, biomedical technology. The technology enables the direct, painless, noninvasive real-time imaging of human nerves and locates points along the nerves that are damaged or inflamed. Though the technology began as a means for investigating chronic pain problems, it will impact the diagnosis and treatment of a large number of diverse neurological conditions. I named the process "Impedance Neurography."

It's important when naming something in science that the name provides information regarding the process. Consider, for example, the Kreb's cycle. Now, most of us who have studied biochemistry know from frequent usage what that means. However, for someone unfamiliar with the underlying biochemical reactions, someone's name tagged on a process does not convey any real information. Contrast that with calling the same process the citric acid, or tricarboxylic acid cycle. Okay, now we have some information regarding what the processes actually involve. The same is true of roentgenogram. What on earth is it? Well, it's a radiogram or X-ray image. Most of us quickly recognize what is meant by an X-ray, but though Roentgen likely would be pleased to be recognized by having his name attached to the images, it's not very helpful. Similarly, Impedance Neurography tells one that an image of a nerve (a neurogram) is being constructed from impedance information. A "Corygram" would not tell anybody anything.

The research and development process we undertook was a classic adventure story, starting with a chance meeting between strangers that demonstrated a curious observation, and just like Bilbo Baggins said, "You step into the Road, and if you don't keep your feet, there is no knowing where you might be swept off to." In this case the road led to a new understanding of nerve biophysics, opening vistas for diagnosis and treatment formerly unavailable. My intent in this work is to achieve three goals:

- The definition of the problem presented by the original observations,
- Presentation of the research path, including twists and turns, that resulted in a new understanding of the underlying biophysical parameters involved, and
- The development of an entirely new technology plus its research and clinical applications.

It is also my intent that this work be accessible to a wide audience and not just practitioners of neuroscience or medicine. This is exciting stuff, a true adventure in which I have been privileged to engage and I hope that some of the excitement and intrigue comes across in recounting the tale. Who knows, it may spur others to a career of such adventures. I have also avoided going into depth concerning some concepts, and this was purposeful on my part for two reasons. (1) To understand concepts and make them one's own, work is required. One simply cannot be told a fact and understand all the nuances related to that information; digging deeper is required. I hope that my bringing up some interesting aspects of electroneurophysiology will spur some people who are so inclined to look more deeply. (2) Some of the information is simply much better covered by other authors, and I have included references to their work.

Now, I would like to introduce the device we constructed to perform Impedance Neurography, shown in Fig. 1.

Fig. 1 depicts a proof-of-concept prototype Impedance Neurography device, the fourth in a line of prototypes, which is connected to a computer via a USB connection. All the electronics unique to the device were contained in the circular portion of the system that had a single switch on the top that turned the device on or off. Control of the output was accomplished via a laptop computer so a large number of different parameters could be assessed using the system. Output waveforms were either square or sinusoidal and user selectable. For safety considerations, the current output of the device was limited to less than 100 μA. That current output limit was more than adequate for the effective functioning of the device without causing any sensation for the subject. That meant its use was painless and noninvasive. The electrode array, made from flex circuitry and attached to the "hockey puck" portion of the device, had 60 electrodes in six rows of 10 electrodes each. The array was attached to the skin surface via a foam interface with hydrogel-filled holes overlying each of the individual electrodes. The foam prevented any hydrogel

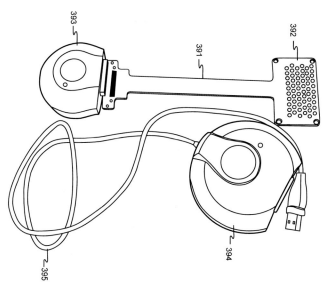

FIGURE 1   Impedance Neurography device. *Disclosed in Cory, US Pat. Appl. 20110082383.*

bridging between electrodes, very effectively controlling the skin contact area. The location and course of underlying nerve structures were determined using point-to-point skin surface electrical impedance determinations.

The device was made to allow flexibility in output control by the computer connection. It's clear that the system could be configured to be handheld, or as small as a disposable liquid crystal display patch, or as large as many hundreds of electrodes in an array to image an area such as the low back. Just think if a surgeon had such a device that could be applied to a surgical site, e.g., the knee prior to arthroscopy, and within seconds of application to the skin visualize where nerves were running under the skin, positioning the incision sites to avoid injuring those nerves. Or a technician performing spine surgery monitoring could simply position electrode arrays allowing the software and hardware to select the best monitoring sites, which may change during the course of the surgery. People suffering motion sickness on airplanes could use a small device similar to the wrist acupressure systems, now available, to electrically stimulate the median nerve-associated site to help control nausea. Such a device would automatically detect the appropriate place to stimulate and could even structure the output to be individualized for the person based on measured parameters of the underlying nerve. More sophisticated stimulation devices, e.g., spinal cord and peripheral nerve stimulators, could be constructed to select best frequencies of stimulation, as well as optimizing other stimulation parameters based on determining nerve electrical characteristics from subsensory, electrical interrogation routines.

All those possible applications, and many others, are enabled by the recognition of how nerves respond to the way an electrical field is applied. This has been heretofore unrecognized, and a basic result of applying this information is that the equivalent circuit model of the nerve cell membrane is changed by a very simple alteration in electrical field parameters. Essentially the nerve membrane equivalent circuit is changed from a parallel RC (Resistance Capacitance) model to a parallel RLC (Resistance Inductance Capacitance) model. For the nonelectrical engineers reading this, that will not make a lot of sense, but it's why I am writing this book—to explain how it is that all those changes occur and why they are so important. This understanding also impacts some of the ways existing technologies function that involve nerve stimulation.

We have not been able to bring this device to market so it's not possible to obtain one. There were several factors, unrelated to the technology itself, responsible for this, but it is my hope that interested parties reading this will be prompted to pick up where we have left off and produce a commercial device.

It is the application of the technology that is really the fascinating part of this story. To that end, I will begin with an illustrative case from my days as a clinician treating chronic pain.

She was in her midforties when I met her and her husband. A Jack Sprat couple, he was thin as a rail and she weighed over 300 pounds. Both smoked incessantly. Sitting side by side, she told her tale and he provided occasional details.

About 20 years earlier, she had been involved in a rear-end collision motor vehicle accident, but that phrase does little to describe what happened. She had stopped her Volkswagen Hatchback in her lane, waiting to make a left turn, when the small, lightweight vehicle was hit from behind by an eighteen-wheeler traveling at approximately 55 mph. The instantaneous forward acceleration was sufficient to cause the driver's seat, into which she was securely seat-belted, to be ripped from its moorings, and when the skidding stopped, she found herself still strapped into the seat, resting against the back hatch door of the Volkswagen. She was lucky to have survived, but the aftermath of the accident was persistent low-back pain that for 20 years had gone undiagnosed as to its origin by several medical and surgical specialists and was unsuccessfully treated with multiple rounds of physical therapy, various medications and potions. Over that time, her weight had steadily increased, and the slightly built girl ballooned into an obese middle-aged woman with a waddling gait. As her size increased, the dismissiveness of her medical consultants increased in kind; depression and despair became part of her persona. Cigarettes and narcotics became her pain relievers of choice to try and reduce her constant discomfort.

My examination revealed that the low-back pain was not a diffuse problem but had a definite area from which it originated. Rather than finding that I could provoke her discomfort from maneuvers directed at the lumbar region, her discomfort clearly came from the left sacral region, just above the buttock. A pain source originating in the low-back or lumbar spine region would have been more expected, and I found her distribution of symptoms to be initially confusing as had many previous examiners.

Reviews of plain X-ray images of her low back were not helpful, as the structures appeared normal. Yet, over the first two or three visits, her complaints were very consistent, always indicating the same left sacral region as the site from which her symptoms emanated. Eventually I proposed to her that she allow me to perform an injection of local anesthetic into the left sacroiliac joint, thinking if the joint was the source of her discomfort, numbing the joint surfaces with the anesthetic might relieve her pain.

As is the case with many desperate people suffering from persistent pain, she readily agreed. Anything that held potential for providing relief was acceptable to her. The difficulty at the time of the procedure was adequately positioning a very long needle into the joint, one of the challenges of working with large individuals. Once proper positioning had been accomplished, an injection of a small amount of local anesthetic resulted in complete pain relief and the answer to her 20 years of discomfort was revealed. Now what?

Though this may seem strange, in pain management this is a frequent question. What does one do therapeutically once the painful anatomy has been determined? What one learns very quickly in pain management is that knowing the source of the discomfort does not automatically point the way to solutions. There are several options in a case such as hers, but the simpler ones had either already been tried or did not provide lasting relief when I tried them. It's also important to remember that this was in the 1990s, and some of the commonly accepted treatments of today were either not available or very new and with a limited track record of use. Since we were stymied at a point where the diagnosis was known and confirmed, but treatment elusive, I asked her if she would allow me to image her back in the area of interest with an experimental Impedance Neurography device I had been developing for several years. This device measured the ease with which electrical current passed through tissue, and I had found was an indicator of the density of nerves in the underlying tissue. The technique will be described in much more detail in a later section.

Since the technique involved was painless and noninvasive, institutional review approval was in place, plus we had developed a good working relationship, she agreed. At the time, performing the procedure was time intensive. The instrument used consisted of a row of seven electrodes with a square of felt, 5 mm on a side, glued to each electrode with conductive glue. The felt was soaked with water, taking care not to squeeze the fluid out which would dehydrate the felt in an unpredictable fashion and then held on the patient's skin while 10 measurements were carefully recorded. Then the process was

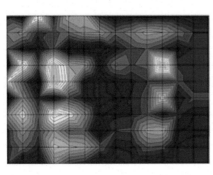 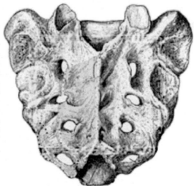

FIGURE 2 Early Impedance Neurography image taken from the sacral region of a patient with low-back pain (left) and diagram of the sacrum (right). *From Bigeleisen, editor.* Ultrasound-guided regional anesthesia and pain medicine, *vol. 42; 2010. p. 278, Fig. 42.3, with permission by Wolters Kluwer Health.*

repeated at the next sampling site. In this case, I sampled two, parallel columns of 10 rows. When I finally was able to graphically display the data obtained from her back, the image depicted in Fig. 2 emerged.

The impedance neurograph on the left side of Fig. 2 showed the electrical parameter of impedance in color with the lighter colors indicating lower impedance. Impedance is similar to electrical resistance; it is a property of material that "impedes" the flow of current and work must be performed to push the current through the material. A more in-depth discussion will be presented in Chapter 1. The two turquoise peaks in Fig. 2 were the lowest impedance values. The squares in the neurograph were 5 mm on a side with the actual impedance values being those seen at the intersections of the square edges. Microsoft Excel was used to construct the neurograph interpolating the impedance values between the intersections. The two turquoise peaks corresponded to the uppermost holes (neuroforamina—literally "nerve holes") in the diagram of the bony sacrum on the right of Fig. 2. The remaining two peaks below each of the turquoise peaks corresponded to the two neuroforamina seen below the topmost ones in the sacrum diagram.

This was a seminal event for me. Looking at the Impedance Neurography image on the left of Fig. 2, it was apparent that there were two, parallel columns of low impedance peaks that corresponded to the neuroforamina of the sacrum. The reason those peaks were imaged was not due to the holes in the bone per se, but to the fact that those holes were stuffed full of nerve tissue. Also, it was apparent that there were low impedance regions to the left of the left-sided neuroforamina; two peaks were seen close to the left edge of the image. This was intriguing since her pain was in the left, sacral region. Those peaks were revealed to be associated with the course of the underlying articular branches (nerves supplying sensation to the sacroiliac joint) of the posterior primary rami of the sacral roots emerging from the neuroforamina. Here is how that was determined.

One of the techniques for identifying nerves that supply a painful structure is to place electrical probes (stimulating needles) in close proximity to the nerves and electrically stimulate those nerves using electrical parameters known to be sufficient to fire nerves, but not to cause any other sensation. If that stimulation reproducibly provokes the individual's pain symptoms, and local anesthetic injection at that same site relieves the pain by blocking the nerve conduction, the nerve supplying the painful part has been identified. That was the case in her situation.

Using the two peaks on the left side of the impedance neurograph as guides for needle insertion, I could provoke her discomfort by stimulating either of the nerves associated with the peaks when the needle tip was just above the surface of the bony sacrum: the position where the articular nerves course over the sacrum to the joint itself. Injection of very small (0.2 mL or one drop) volumes of local anesthetic at one or the other of the two nerves associated with those peaks did *not* result in appreciable pain relief, but injection of the same, small volumes of local anesthetic at *both* sites, simultaneously, did result in pain relief. The importance of injecting small volumes of local anesthetic was that the medication from one injection would not spread far enough to cause effects at both nerves; it was only effective at the nerve being stimulated electrically.

This finding was reproducible on two separate occasions, and a third session was scheduled to perform radiofrequency ablations of the two articular nerves. Radiofrequency ablation disrupts nerve functioning by local heating that is accomplished with needle electrodes under local anesthesia. When well done, with attention to providing adequate anesthesia with local anesthetic, it is painless for the patient on whom the procedure is being performed. Also, it is temporary in its effects so, over time, recovery of normal nerve function will occur. The question is open until that recovery has happened as to whether the pain will return. Following the ablation procedure, she had 5 years of good quality relief from her low-back pain.

There was another aspect to this procedure that may not be obvious. Recall that she weighed in excess of 300 pounds. The location of the articular nerves, which are typically about 2 mm in diameter, was 8 cm below the skin surface, yet using Impedance Neurography, for guidance, it was possible to place the needles in proximity to those nerves on the first pass of the needles. This is not possible using X-ray guidance for needle placement as the nerves are not visible via X-ray and are too small for detection by MRI or ultrasound. Plus, at an 8-cm depth, the nerves were beyond ultrasound imaging limits. This was, for many years, an enduring mystery about Impedance Neurography; how was the technology able to accurately image those small nerves at such a great depth? And why only on the left? Looking at the impedance neurograph in

Fig. 2, there were no corresponding peaks on the right edge as there were on the left though articular nerves also coursed over the surface of the right sacrum supplying the right sacroiliac joint. I now know the answer to those questions.

The story was bittersweet, however. I closed my pain practice shortly after performing the ablation procedures and lost contact with the couple. Several years later, I received a phone call when in the operating room performing surgical anesthesia. The caller was her husband who related that they had subsequently moved out of the area and when her low-back pain eventually returned, she was unable to obtain repeat ablation procedures despite seeking help from different providers; the unremitting pain wore her down. Depression once again set in, and having lost all hope, she eventually committed suicide. Today, radiofrequency ablations of the sacral articular branches are a relatively common pain management procedure, but then they were largely unknown. The results of her inability to identify another provider willing to undertake a very effective procedure to provide pain control were tragic, but all too common.

As I wrote this work, I searched online to see if I could find any traces of her. After quite a bit of looking I found a death notice in the online Social Security records. There was no online obituary, and the Social Security record was all I could find. Fifty years of life, and it came down to a few lines of print with a date and location of birth and a date and location of death: nothing else. Her life reflected the fact that we all too often overlook or disregard people for various reasons. Due to three, common factors of gender, obesity, and chronic pain, her situation and her interaction with medical professionals exemplified Rem Edwards' comments in his paper, Pain and the Ethics of Pain Management,

> *Challenging the patient's claims concerning the degree of his suffering and/or his need for help definitely and gravely affects his moral standing in the hospital and the broader human community. Such a person is tacitly branded as a liar, and he also comes to be regarded as somehow cowardly, uncooperative and lacking in will power or attention power. Consequently, he is relegated to the inferior status of being a defective moral agent.*[1]

I think about her, occasionally, and now wonder if anyone else ever does so especially since the couple had no children. It is my privilege to pen a few lines about her as someone who played an important role in the whole discovery of the underlying biophysics of neurons enabling the technology of Impedance Neurography. Though she used the world lightly and left no evidence of her passing, she played a valuable role in something of worth.

There was an additional, intriguing observation, consistently seen over the years of my examining many people with painful conditions. A hierarchy of nerve structures was visualized by Impedance Neurography. No other tissue types, such as vessels or muscles, were ever imaged with this technology. Normal nerves were clearly seen on impedance neurographs as low impedance, linear structures. Even lower impedances were seen from nerve branch points, which often corresponded to classic acupuncture sites or muscle abnormalities called myofascial trigger points. Entrapments and nerve contusions showed lower impedances yet, probably related to variable degrees of demyelination associated with those conditions, but it was neuromas (damaged nerves that have attempted to regrow) that demonstrated the lowest impedance measurements. I observed that this hierarchy corresponded with the amount of exposed neuronal cell membrane associated with the structures, but in the early years of researching the effect did not understand why. It was an important clue that will be explained later in this work.

REFERENCE

1.   Edwards RB. Pain and the ethics of pain management. *Soc Sci Med* 1986;**18**(6):515–23.

Chapter 1

Initial Impedance Neurography Findings: The TENS Technique and Nerve Stimulation Observations

A restatement of Bertil Hille's comment from the 1977 Handbook of Physiology is as follows:

> *Current has* nothing *directly to do with nerve stimulation.*
>
> Hille [1]

Neuroscientists have known this for well over 60 years, though the message does not seem to have filtered down to the clinical side the way it should. I use the term "directly" because the current is associated with a driving voltage, a factor that will be discussed as I go along. But first, let us turn to some interesting observations I was shown in the mid-1980s. These observations started with finding a muscle abnormality called a myofascial trigger point. The observations ended up not only providing a more complete understanding of how nerves are stimulated by applying an electric field in the vicinity of the nerve but also led to a new way of imaging nerves. I will use the term "externally applied electric fields" to describe this approach. It's actually no different from applying electric fields directly to the nerve cell membrane but involves additional sources of impedance besides that of the membrane. That difference has been a source of some confusion over the decades, which I will explain before getting to the actual observations.

The reason for confusion relates to the structures traversed by the current when applied from different locations. In most single neuron studies, the electrodes are attached to the neuronal membrane and typically spaced around 200 microns apart, or one is attached on the external surface of the neuronal membrane and the other located opposite on the inside of the membrane. In either of these configurations, a discrete patch of neuronal membrane represents the resistance or impedance to current flow. Being discrete, that patch of membrane can be treated as having fixed electrical parameters of resistance and capacitance. Consequently, a current can be applied to that patch of membrane and it will be associated with a fixed, driving voltage. If the current

Finding the Nerve. http://dx.doi.org/10.1016/B978-0-12-814176-2.00001-0

is time variant, e.g., alternating current or AC, the voltage will vary depending on the frequency of the current waveform, but for any given frequency, a particular current will be associated with a particular voltage. In other words, the patch of membrane can be treated as a predictable, electrical component.

Now a bit of history comes into play.

Prior to the 20th century, and the invention of vacuum tubes in 1904, the way to construct an electrical power source was as a controlled current source. All this required was a battery (or pile as they were called) and a resistor. These conventions continued into the 20th century in studies of electrophysiology such as described in the above paragraphs, and constant current became dogma in studies of the neuronal cell membrane. This dogma is reflected in a conversation a colleague had with his mentor in his postgraduate neurophysiology studies. My colleague was setting up a system to electrically stimulate isolated neurons and he was confused as to why the good practice in nerve stimulation design included a provision for constant current. He asked his mentor why he should use constant current and reported that the reply was, "Because I'll kill you if you don't." Not particularly satisfying as a response, but his mentor got the point across. In researching this issue, it appeared that the use of constant current in nerve stimulation studies is rarely questioned.

Mainly this was a measurement issue, early on, as voltage can easily be measured whereas the direct measurement of current is more difficult. Consequently, an electrophysiologist could apply a constant current to a patch of cell membrane, measure the resulting voltage, and calculate the resistance or impedance of the membrane. Easy!

If a constant current is applied to a patch of nerve membrane at an amplitude that causes the nerve to depolarize or fire, that current can be reproducibly associated with the ability to fire the nerve reflecting the notion of the fixed electrical parameters of the patch of membrane. The question that remained unanswered was whether the current actually caused the nerve to fire or whether the voltage associated with driving the current across the fixed resistance or impedance of the membrane was the important factor. The problem was complicated by findings in the last part of the 19th century that potassium solutions applied to a nerve surface would cause it to fire. So, what was it about potassium ions?

At the time of the finding related to potassium solutions and nerves, the notion of ions migrating across a membrane depending on their concentrations on each side of the membrane was understood. Thus, the idea developed of the importance of the concentration of ionic charges lined up across the membrane that constituted the main factor leading to depolarization. Current is defined as movement of charge, and as a consequence of the observations related to potassium ions, current and the resultant charge relocation became the assumed critical factor in nerve depolarization. The unrecognized problem was the observations derived from a small patch of neuronal cell membrane and its relatively fixed resistance or impedance. This meant that whenever a single current pulse was applied to the membrane, the resistance or impedance was constant, so the relationship appeared to be between current amplitude, current pulse duration, and depolarization. What was not clear was what happened when the current supplying electrode was remote from the nerve: an externally applied field. In that situation, there exist multiple resistance and impedance structures so the current–voltage relationships at the nerve cell membrane were not possible to predict as many more structures than a discrete patch of nerve cell membrane were involved in the current path, each with their own current–voltage relationships that contributed to the whole. It was not until the mathematics of neuronal cell membrane depolarization were worked out that the error of considering current of primacy in nerve membrane depolarization became obvious, hence Hille's comment at the beginning of this chapter.

Let us get back to the Impedance Neurography journey...

The first observation that led to the realization that nerve-rich tissue was identifiable by the application of periodic electrical fields involved probing for myofascial trigger points with a Transcutaneous Electrical Nerve Stimulation (TENS) device. This was shown to me by a physical therapist who had become a medical device representative. I had never heard of it prior to that time, a fact that seemed quite surprising to him. TENS units usually employ pulsed direct current waveforms. The waveform may be monophasic or biphasic (Fig. 1.1) with adjustable frequency and pulse width.

The technique using a TENS unit to identify a myofascial trigger point involved placing one of its leads on the subject remote from the site of interest and placing the other lead in the examiner's hand. Then the examiner would put some conductive gel on his/her index finger of the hand holding the TENS unit lead and touch that finger to the subject slightly away from the area of interest. The TENS unit amplitude was increased until the examiner felt a pronounced (!) electrical shock sensation in the index finger, but a really interesting finding was that the subject felt no sensation at any point during the procedure. The finger was then moved back and forth over the suspected site of

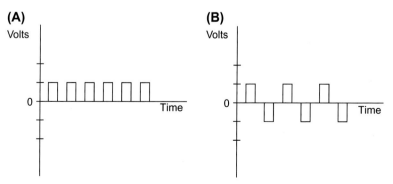

FIGURE 1.1 A monophasic positive square waveform is seen in example (A). Example (B) demonstrates a biphasic square waveform.

the myofascial trigger point, and the electrical shock sensation was noted to increase and then decrease as the finger crossed the myofascial trigger point. Additional checking around the trigger point revealed a line of increased stimulation sensation that intersected the trigger point. This line was defined by crisscrossing it with the probing finger in the fashion described above. Increased electrical shock sensation occurred when crossing the line, although to a lesser degree than that of the site overlying the myofascial trigger point.

Fig. 1.2 shows the setup of the TENS technique over a peripheral nerve. The TENS unit connected the subject and the examiner with the examiner's finger completing the electrical circuit. This same technique may be used to identify acupuncture points and is described elsewhere in somewhat mystical terms as, "Chasing the Dragon." [2,3] Those having TENS units should try this themselves since experiencing the effect is everything. It's also a difficult observation to quantify; after all, how does one accurately measure the difference in sensation of electric shock intensity? And, by the way, you cannot do it on yourself. If you think about the electrical circuit, it's obvious that having two leads on one subject results in a complete circuit without the need for an intervening finger. A subject is required to demonstrate the effect. It's a very subjective observation and one cannot design equipment to objectify what one feels, but the observation is so important.

Further investigation demonstrated that a stimulating needle for regional anesthesia advanced through the skin at the site overlying the trigger point, and on a normal (90 degrees insertion angle) to the plane of the skin surface,

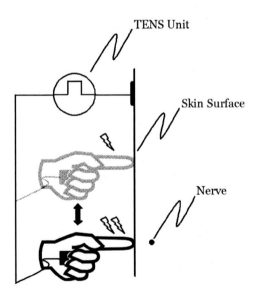

FIGURE 1.2   Diagram of the Transcutaneous Electrical Nerve Stimulation (TENS) technique setup with TENS unit electrodes attached to the subject and the examiner's hand.

reproducibly intersected the myofascial trigger point. When the stimulator was turned on with the needle at the myofascial trigger point, the muscle containing the trigger point twitched. Notably, those myofascial trigger points were *always* found on the muscle side of the deep fascia of the involved muscle, while acupuncture points could correspond with myofascial trigger points or be associated with subcutaneous nerve branch points. The anatomic nature of myofascial trigger points is discussed later in a few paragraphs as well as in Bigeleisen.[4]

The same effect observed at the myofascial trigger point was also observed if the stimulating needle was inserted proximal to the trigger point along the line of increased electrical shock sensation that was often observed intersecting the trigger point. A normal insertion angle was invariably required along that line to find a site where stimulation resulted in a twitch of the muscle involved in the myofascial syndrome, and sometimes muscles further distal to the stimulation site were also observed to twitch. The conclusion these findings most suggested was that the line mapped by the increased electric shock sensation that intersected the trigger point corresponded to the course of a nerve, and for some reason the nerve could be detected at the skin surface using a TENS unit as described. These nerves could be surprisingly deep, often in the range of 4−5 cm below the skin surface, but always lying on a normal to the curve of the skin surface at the site of maximal increased electrical shock sensation. No literature offered an explanation for the observation; however, with early prototype instrumentation the skin surface sites were found to correlate with a regional decrease in skin surface impedance. This change in skin surface impedance has been noted by some, but not all investigators.[5,6] A case story illuminating the mystery of myofascial trigger points follows.

IMPEDANCE NEUROGRAPHY REVEALS MYOFASCIAL TRIGGER POINTS

One day, a patient presented to my clinic with left upper back pain. His story was that he had been working with a partner on loading a trailer with materials for a job site. The trailer was not hitched to their pickup so my patient had placed one foot on each side of the hitch and bent over to pick it up. Unbeknownst to him, at the same time as he was bending to lift the hitch, his partner pushed down on the rear end of the trailer to counterbalance it and make the hitch easier to move, but he overestimated how much counterbalancing was needed. His weight was sufficient to lever the hitch up very rapidly, out of my patient's grip. The hitch struck my patient under his chin, snapping his head back and knocking him onto the ground. He stated that lying there, he felt as though he was unable to move his body (not a good sign…). However, his partner jovially came around to the front of the trailer, amused at what had happened, grabbed him by the wrists, and yanked him to his feet. During that maneuver, my patient stated he felt a lightning-like bolt course down from his neck to his feet and found himself able stand, unaided, with all body sensation apparently normal. To this day, I have absolutely no idea what happened during his actual injury event, but at least it seemed to work out okay. And, though it could have indeed been worse, it was not actually good; he ended up with a persistently painful neck and left posterior shoulder discomfort. Physical examinations and imaging studies including planar X-rays and CT scans had been unrevealing as to the nature of his injury and he came to me seeking help with the posterior shoulder pain.

Examination revealed that his posterior shoulder pain was entirely muscular in origin, constituting what is termed a "myofascial syndrome" with multiple myofascial trigger points present. Those sites were revealed by palpating the muscle beneath the skin and discovering tender nodules within the muscle. Such trigger points have been recognized literally for centuries as they overlap substantially with the position of classic acupuncture points. Over several visits, it became apparent that there was one, primary trigger point with many satellite trigger points, which is not an uncommon situation in clinical practice. Injection of local anesthetic at the primary trigger point resulted in resolution of the satellite trigger points, but the primary site would return after a few days.

His situation afforded the opportunity to use Impedance Neurography to image a persistent myofascial trigger point, to which he gave his consent. The impedance neurograph of his trigger point is shown in Fig. 1.3 using the same device and protocol as described in the Introduction.

An obvious impedance minima was seen at row 2.5, column S6 on the right side of the image. This site corresponded to a point on the skin surface overlaying a nodule in an underlying muscle that was tender to palpation: the definition of a myofascial trigger point. Inserting a needle at this site and directing it on a normal to the skin surface during advancement revealed the trigger point site to be on the muscle side of the deep fascia of the levator scapulae muscle. This is the "That's it!" site; a classic way of identifying trigger points performed by advancing the needle until the patient says, "That's it!". Patients are *very* definite about those sites.

As an aside, injection of trigger points involves some art and the use of appropriate equipment. I used to tell my residents when teaching them how to inject medication, including for treating trigger points, that one must listen to their fingers when performing injections. The tip of a needle will encounter multiple tissue structures as it's advanced toward the intended target and perceiving those tissues is key to effectively performing the injection. For this reason, it's important to use a 45 degrees bevel needle that provides a bit more resistance to advancement, particularly through fascial tissue planes. Feeling the increased resistance and the "pop" as the needle tip breaches the tissue plane is important for proper procedure performance. Having said that, most clinicians who purport to inject trigger points use 35 degrees bevel needles that do not supply effective feedback sensation. Their rationale that a longer bevel

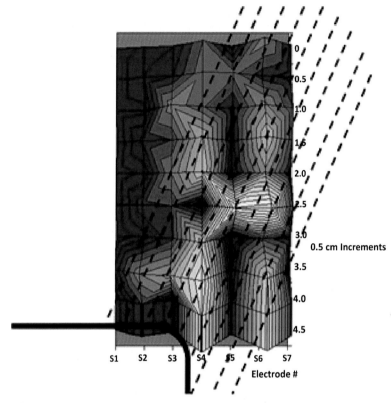

FIGURE 1.3 Impedance neurography of a persistent myofascial trigger point. The heavy, bold line at the bottom left represents the superior medial border of the left scapula, while the *dashed lines* represent the course of the levator scapulae muscle. *(From Cory P, Bigeleisen P. Impedance neurography. In: Bigeleisen P, editor. Ultrasound-guided regional anesthesia and pain medicine. Philadelphia: Wolters Kluwer|Lippincott Williams & Wilkins; 2010;42:278, Fig. 42.4, with permission by Wolters Kluwer Health)*

needle is more comfortable to use simply reflects their haste in procedure performance because they do not care to take the time to preinject local anesthetic. Clinician convenience is never a reason to avoid using correct equipment or technique.

Consider the case being discussed. The involved muscle, the levator scapulae muscle, lies deep to the trapezius muscle. Most clinicians on palpating a nodule medial to the upper, medial corner of the scapula will pronounce that the patient has a trigger point in the trapezius muscle with no confirming evidence, grab a 5/8″ long needle (much too short to reach the target), and inject into the substance of the trapezius without any indication that the "that's it" spot has been identified while announcing that they have injected the trigger point. As a consequence, there is no benefit and the result is that the literature is replete with studies reporting mixed results for trigger point injection therapy. It's not that trigger points do not respond to well-targeted injections, but they certainly do not respond to injections that miss the target. To accurately identify that the trigger point in question lay in the levator scapulae muscle, the fascial pops encountered during needle advancement must be counted. There will be one pop for the superficial trapezius fascia, a second pop for the deep trapezius fascia, a third for the superficial fascia of the levator scapulae, and a fourth for the deep fascia of the levator scapulae. In the case being presented, I counted three pops without any feedback from the patient that would indicate I was close to the "that's it" site. But when the needle bumped into the fourth fascial boundary, he was quick to state I had reached the site.

At that point, I engaged in some additional investigation. Using a nerve stimulating needle, when stimulating at the site of bumping into the deep muscle fascia, the levator scapulae muscle twitched indicating that the needle tip was in proximity to a mixed function nerve (mixed motor/sensory function) supplying the muscle itself. On reinspection of the impedance neurograph in Fig. 1.3, it's apparent that a line of decreased impedance runs through the peak described above. The question was, did that line represent nerve tissue?

Advancing the stimulating needle just through the deep fascia of the levator scapulae and stimulating again, the same twitch of the levator scapulae was observed, and a distal muscle, the rhomboideus major, also twitched. That result indicated that the needle tip was now in proximity to a nerve supplying both muscles: the dorsal scapular nerve. Repositioning the needle on the muscle side of the deep fascia and stimulating, once again resulted in a twitch of the levator scapulae muscle, but not the more distal rhomboideus major. What was clear from this result was that the myofascial trigger point was associated with a branch point where a mixed function nerve supplying the levator scapulae muscle branched from the larger dorsal scapular nerve. This association with a vertical branch point is interesting in light of subcutaneous acupuncture points being found to correlate with horizontal branch points of subcutaneous nerves. As stated previously, many classic acupuncture points overlap anatomically with the position of known myofascial trigger points, and nerve branching appears to be a common feature.

Injection treatment for myofascial trigger points involves using local anesthetic followed by physical therapy to try and restore normal muscle function. Techniques mixing corticosteroids with local anesthetic do not provide any additional benefit, and "dry needling" is ineffectual and inadvisable since it often causes scarring and worse symptoms over time. Worst of all, the multiple needle passes used in dry needling can injure or transect the mixed function nerve involved in the trigger point, effectively converting an often eminently treatable condition, the trigger point, to a much more difficult management situation involving a posttraumatic neuroma formed as the damaged nerve attempted to repair or regrow. However, local anesthetic combined with physical therapy sometimes is not enough and a useful technique is to inject a very small amount, e.g., 10 ng, of botulinum toxin (Botox), on the involved mixed function nerve. The botulinum toxin enters the nerve where it works peripherally to prevent neurotransmitter release and is also transported via fast axonal transport centrally to block neurotransmitter release where the nerve enters the spinal cord. This technique can provide months of good quality relief from persistent myofascial trigger points. To be effective, it's important that the botulinum toxin is injected in close proximity to the very small, mixed function nerve involved in the myofascial trigger point. Without marked patience on the clinician's part in positioning the needle, or an effective targeting method such as two-dimensional trigger point mapping with an Impedance Neurography device coupled with nerve stimulation to identify the nerve in the third dimension of depth, efficient treatment of trigger points with botulinum toxin is not possible.

Botulinum toxin injection was performed at this trigger point and my patient received several weeks of good relief. He then called to tell me the trigger point had returned, so reimaging was performed resulting in Fig. 1.4.

Comparing Figs. 1.3 and 1.4 demonstrates the reproducibility of Impedance Neurography for therapy follow-up and documentation. Note in Fig. 1.4 that the peak seen in Fig. 1.3 is gone, but a new peak has emerged at row 2.0, column S4. The botulinum toxin effect on the peak in Fig. 1.3 appeared to be persisting, but a new myofascial trigger point had developed. Note, also, that in both Figs. 1.3 and 1.4 there are two, parallel lines running from top to bottom of the impedance neurograph in columns S4 and S6. It turned out that repeating the investigative steps from those performed on the peak in Fig. 1.3 to the peak shown in Fig. 1.4 gave the same results. It appeared that the two, parallel lines represented a reduplicated left dorsal scapular nerve in this patient. Repetition of the botulinum toxin injection at the new trigger point resulted in several months of good quality pain relief.

This is actually the first demonstration of the nature of myofascial trigger points. For decades, these have been thought to be muscular problems associated with palpable, tender nodules in the muscle, but those nodules are downstream effects of the primary site of action at the myofascial boundary. It appears that myofascial trigger points are entrapment phenomena wherein mixed function branching nerve fibers are entrapped in the investing fascia of muscles, always on the deep muscle/fascia boundary. This anatomy makes sense because the nerves supplying muscles run underneath the muscles, sending multiple branches up into the muscle substance. These branching fibers are not purely motor nerves, as often stated in the literature. On microscopic examination, upward of 40% of the axons contained within those nerves are small, unmyelinated sensory fibers that likely explain how it is that myofascial trigger points can be so uncomfortable. The appearance of the myofascial boundary entrapment has yet to be studied, since up until now no literature has pointed to those sites as the significant pathologic location. However, the relationship to the muscle/fascia boundary is a finding that is not unique to me and is reflected in the name: myofascial trigger points.

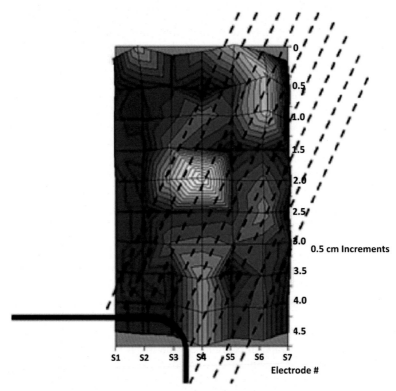

FIGURE 1.4 Repeat impedance neurography of the region shown in Fig. 1.3 at 3 months postbotulinum toxin injection. *(From Cory P, Bigeleisen P. Impedance neurography. In: Bigeleisen P, editor. Ultrasound-guided regional anesthesia and pain medicine. Philadelphia: Wolters Kluwer|Lippincott Williams & Wilkins; 2010; 42:278, Fig. 42.3, with permission by Wolters Kluwer Health)*

I recently followed up with this patient who informed me that over 20 years after the fact, his shoulder pain has remained well managed. This is a fascinating result considering that I thought his myofascial syndrome was being driven by an injury to his cervical spine and treating the myofascial trigger points would be at least one step removed from the primary process. Instead, he has derived long-term relief from highly targeted therapy directed at the nerve entrapment in the myofascial boundary. The story brings up a couple of observations.

The first observation is the old nugget that absence of evidence is not evidence of absence. We are increasingly seeing the error of automation bias in medicine where laboratory and imaging studies are being given primacy over clinical acumen. In the examples discussed in the Introduction and immediately above, all the imaging studies were normal and yet the two individuals were suffering significant discomfort. It was only by believing their stories and being willing to ignore hypothesized diagnoses of convenience, e.g., conversion reaction, that the true nature of their conditions was eventually revealed. I have often told medical students and residents that if they are making the diagnosis of conversion reaction or somatoform pain disorder, they are most likely wrong. I have yet to see that admonition proved incorrect.

The second salient observation is that to successfully ferret out more difficult diagnoses, two critical characteristics are required of the physician: curiosity and persistence. In today's climate of medical practice, these two factors are difficult to maintain due to productivity pressure from employers, reimbursement pressure from insurers, and excessive record-keeping burdens imposed by electronic medical record systems. Then there is the issue that new physicians are trained to get the "right" answer rather than being encouraged to be curious about the situations that do not exactly fit in any particular category. Sometimes physicians need that persistence, or patience, to begin to understand what they are seeing. One of the main drivers to developing instrumentation that accomplished what I was able to perceive with the TENS technique was that other physicians were seemingly unable to do it. On watching them try, it was obvious that they were simply too impatient to perceive much of anything in their fingertips. If the observation was not immediately obvious, they were done trying which was something I found quite disheartening. Regarding curiosity, the most dramatic instance I can recall was when I tried to demonstrate and explain to a colleague about the science that underpinned our technology. He stopped me, midsentence, to say, "Phil, I don't care *how* it works — I only care *that* it works." I am not certain I have any answers to all of this, but we do need to think about where we are going.

Continuing on…

CONFUSING RESULTS REGARDING THE ROLES OF VOLTAGE AND CURRENT IN EFFECTING ACTION POTENTIAL DEVELOPMENT

At this point the discussion becomes a bit more technical. I will try to illustrate the points as I make a case for neither current, nor charge, nor applied voltage being the important factors in causing nerves to fire, a process known as depolarization. Math is involved in the analysis and is effective as a shorthand describing some complex processes that I will try to explain in prose as well.

The finger electric shock findings were not only useful in identifying pathologic structures, e.g., trigger points and neuromas, but also provided deductive insight into the nature of nerve stimulation with externally applied electrical fields. The observation of increased electric shock sensation in the examiner's finger when traversing a region of decreased impedance may not seem to be a problem initially, but on reflection, this was difficult to explain as just the opposite result, that of decreased shock sensation, might be expected. Importantly, the effect was seen using both controlled voltage and controlled current stimulators. Though not precisely correct when dealing with impedance, Ohm's Law (Eq. 1.1) is helpful in elucidating this observation:

$$E = IR \qquad (1.1)$$

where E is electromotive force (emf) in Volts, I is current in Amperes, and R is resistance in Ohms. For purists, Ohm's Law related to tissue should be written as shown in Eq. (1.2):

$$E = I|Z|e^{j\phi} \qquad (1.2)$$

where $|Z|e^{j\phi}$ is called the equivalent impedance (Z_{eq}), with j the square root of -1 and ϕ the phase angle. The important thing to understand is that both R and Z_{eq} are defined by the E/I ratio, i.e., they are determined by how much current flows in response to an applied voltage across the resistance (or impedance). If either R or Z_{eq} decreases, the amount of voltage required to drive a fixed current will also decrease, or conversely, the amount of current flowing in response to a fixed voltage will increase. Though not commonly discussed, it is worth remembering that both resistance and impedance are calculated quantities that have no meaning without knowing the voltage required to drive a particular current.

Consider the situation of controlled voltage output where E is fixed at a specific value. If R decreases (over a trigger point), current flow (I) will increase in response to the fixed voltage (E), i.e., the same force (Volts) will drive more current across a decreased resistance. One interpretation of this finding is that the increased current flow through the examiner's finger accounted for the increase in electric shock sensation. However, most TENS units use controlled current with I fixed at a particular value that is user selectable. If R falls, E will also fall as it takes less voltage to drive the same amount of current through the lower resistance, but the shock sensation was noted to increase in the examiner's finger with no change in the current and a fall in the applied voltage. In other words, when using either controlled current or controlled voltage nerve stimulators, the results noted in the examiner's finger were the same.

So, now we have a conundrum. In one situation (controlled current), the voltage falls over a low impedance site, the current remains the same, and the electric shock sensation increases. In the other situation (controlled voltage), the voltage stays the same over a low impedance site, the current increases, and the electric shock sensation increases. If the increased shock sensation was due to increased current flow (second situation), then when the current flow remains unchanged (first situation) and the applied voltage falls, one would at least expect the shock sensation to be unchanged or more likely to decrease, but it increased. How can it be both ways?

It turns out that there is a common feature of both the controlled current and the controlled voltage situations, which is that an electric field is generated. In either a controlled voltage or controlled current situation, an electric field in Volts per unit distance is created. The voltage difference between two points in the field determines the current that flows through the material in which the electric field has been created. The observations from using either controlled current or controlled voltage indicate that firing of a nerve is not related to the absolute current flow, an understanding that has been well-documented for decades, nor with local charge densities, nor with absolute voltage. As Hille states, "The conductance changes apparently depend only on voltage and not on the concentrations of Na+ or K+ or on the direction or magnitude of current flow."[1] By "voltage," what Hille means is the transmembrane voltage gradient in Volts/unit distance.

If the voltage measured at point "x" is 100 V and 1 cm away the voltage is also 100 V, there is no gradient. This means that even though an absolute voltage exists, since it does not change over the distance involved, no current would flow between the two points. In other words, the absolute voltage does not drive current flow. If the voltage measured 1 cm from point "x" is 99 V, then a voltage difference is present and current flows between the two points because charges (electrons, ions, or charged regions of molecules) are compelled to move by the force of the gradient (1 V/cm in this example).

We have known since the 1950s that sodium channel opening (required for action potential generation) is caused by a voltage difference across the neuronal cell membrane. This is reflected in the name of the channels: voltage-gated sodium and potassium channels. The name implies that a voltage differential across the channel is required to open the channel. Once this voltage difference is established, movement of the positively charged portions of amino acid moieties, i.e., arginine and lysine residues, occurs in the voltage-sensing, fourth membrane-spanning segment of each of the four protein repeats of the sodium channel. The charges move in response to a force applied to them, a voltage gradient, which is the definition of an electric field: Volts/unit distance. Here is the interesting question; what happens to the electric field in the TENS technique?

GETTING BACK TO CURRENT AND VOLTAGE BASICS

There is a stumbling block commonly encountered when considering voltage and current relationships required for nerve stimulation. The difficult concept to grapple is that the establishment of a transmembrane potential gradient capable opening sodium channels, by an external electrical field, is not dependent on the amount of current flowing. The start of studies leading to this conclusion begins with jewelry.

Centuries ago, in ancient Greece, a problem was noted with amber jewelry; small fibers would stick to the amber and rubbing only seemed to make the fibers stick more effectively. The observation was known as the "amber effect," and philosophers at the time wondered what was going on that explained the fuzzy pendant problem. Reflecting the role played by the Greek philosophers, we have kept the Greek word for amber, "elektron," as part of our modern scientific lexicon. It required the studies of electricity of Benjamin Franklin to clearly delineate the reason. In the mid-1700s, among other groundbreaking discoveries that Franklin made regarding electricity (commonly overlooked because of his important political activities), he determined the notion of charges as positive and negative, noting that positive charge was that left on a glass rod when rubbed with silk and negative charge was found on a hard, rubber rod when rubbed with cat's fur. He had arbitrarily guessed correctly in his assignment, as was shown much later. Knowing about the existence of charges facilitated other studies, such as those of Charles Coulomb.

Charles-Augustin de Coulomb determined in 1785 that fixed sources of charge in space exert a force on one another that is inversely proportional to the square of the distance between the charges as in Eq. (1.3):

$$F = K\left(Q_1 Q_2 / d^2\right) \tag{1.3}$$

where F is the force in Newtons, Q_1 and Q_2 are charges in Coulombs, d is the distance between the charges, and K is a proportionality constant equal to $1/4\pi\varepsilon_0$. The dimensional constant, ε_0, is termed the permittivity of free space. This relationship reveals that an electric field of 1 V/m will exert a force of 1 N on a charge of 1 C. We often think of this as opposite charges attract and like charges repel. So, the force experienced by two, oppositely charged entities will be such that the entities are attracted toward each other, whereas two like charged entities will experience a force pushing them apart. If you want to see this effect in action, take a small piece of aluminum foil and tie a thread around it. Then, rub a plastic object, e.g., a comb, with wool, on the carpet, or something similar. Slowly bring the charged comb close to the foil suspended by the string and it will defy gravity and move toward the comb. As long as the comb does not touch the foil, the string will not hang straight down, but will be displaced toward the charged plastic object. What you will be demonstrating is an example of Coulomb's Law.

Coulomb's Law is an electrostatic statement that has nothing to do with the electrodynamic situation of nerve stimulation, though it is often trotted out, inappropriately, to explain distance relationships for stimulation. Rather, the Law describes how stationary or immobile charges interact by developing force/unit distance, i.e., an electric field. This field describes the force experienced by the two opposite charges and how that force, measured in Volts per unit distance, varies over the space surrounding the charges. If additional mobile charges, which are not fixed in space, exist within that electrical field, they migrate toward the fixed charge of opposite sign. Thus, in a field between a fixed negative charge and a fixed positive charge, an additional, mobile positive charge will move toward the fixed negative charge and a mobile negative charge will move toward the fixed positive charge. This charge movement in an electric field constitutes current flow, but an important concept to understand is that the presence of the field does not depend on charge movement, i.e., *a voltage field may exist without current flow, but current flow requires the presence of a voltage gradient* (excepting superconductivity—not commonly found in tissue).

Another important concept is that at various points within an electric field, current levels or voltage levels will be the same. I will use the conventional terms isocurrent to describe the situation where current is constant and equipotential to denote equal voltage values.

The notion of isocurrent, current being the same at two places in an electric field, can be illustrated by thinking about a wire with current flowing through it. If the current is measured at either end of the wire, it is the same; "x" Amps enters the wire at one end and "x" Amps leaves the wire at the other end. This situation is one of isocurrent. However, when the voltage is measured at both ends and compared, the values are different reflecting the work being done to "push" charges through the wire (current flow), which is reflected as heat generation. This one-dimensional example of a wire can be expanded to a three-dimensional block of material as shown in Fig. 1.5.

The direction of current flow is at right angles to the equipotential lines of force in the electric field depicted in Fig. 1.5. In other words, the isocurrent lines are always at right angles to the equipotential lines of force. In a block of homogeneous material, the equipotential lines of force and the percentage current flow (isocurrent lines) are shown in Fig. 1.5.

When one traces along a single isocurrent line in Fig. 1.5 many equipotential lines of force are crossed so that at any two points on an isocurrent line

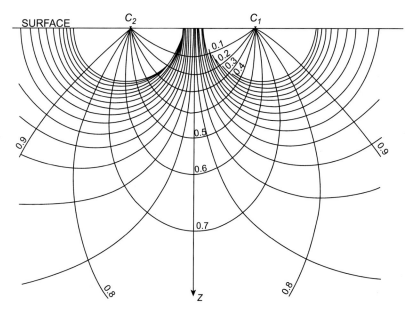

FIGURE 1.5 A demonstration of the distribution of isocurrent lines flowing between two surface point sources and the associated equipotential lines of force driving the current.[7]

the voltage per unit distance is different. Notice in Fig. 1.5 that the isocurrent lines are labeled as decimal amounts. This reflects that each line corresponds to a percentage of the total current flow. Likewise, tracing along a single equipotential line results in crossing several isocurrent lines. Analogously to the isocurrent example, at any two points along an equipotential line, the current is different. Again, at some points in the field the current levels are equal and at others the voltage levels are equal.

The applied voltage (V) in Fig. 1.5 is distributed between the two electrodes (C_1 and C_2) as (+) V/2 and (−) V/2. If one applies 10 V between the two electrodes, the voltages at the electrodes, measured from the field midpoint, are (+)5 V and (−)5 V for a 10 V differential. The equipotential line with 0 V/distance in Fig. 1.5 lies half way between the two electrodes since the bulk conductor in the example is homogeneous, effectively dividing the conductor into two halves of equal impedance.

ELECTRIC FIELDS IN NONHOMOGENOUS MATERIAL

Fig. 1.5 represents an idealized situation, but life is seldom ideal and the TENS technique is illustrative of that fact. Referring back to Fig. 1.5, if one half of the material undergoes an impedance reduction what happens to the 0 V/distance equipotential line that is marked with "Z"? In this situation, the 0 V/distance equipotential line will shift *away* from the lower impedance portion into the higher impedance portion. Why is this the case? The 0 V/distance equipotential line defines the boundary where the impedances on both sides of the line (or plane in three dimensions) are equal. For this to happen when impedances are unequal on each side of the original position of the 0 V/distance line, more material must be added to the lower impedance side and subtracted from the higher impedance side, which moves the 0 V/distance line into the higher impedance side.

With the shift of the 0 V/distance line, the spacing between one of the electrodes supplying the voltage and the 0 V/distance line is reduced, while the spacing is increased for the other electrode. While the distance between the electrodes and the 0 V/distance line changes, the voltage differential from the electrodes to the 0 V/distance line stays the same (0.5 V in this example). Thus, the equipotential line spacing becomes smaller on the higher impedance side, which is another way of saying that a steeper voltage gradient exists. This spacing change of the equipotential lines of force accounts for the sensation change in the probing finger using the TENS unit technique because larger voltage gradients are being created. Notably, all the current flows through the examiner's finger, no matter the total impedance of the system. Due to the dimensions involved, the examiner's finger is the higher impedance component of the system. This is something that will be covered in more detail later. Though the absolute applied voltage has not changed in the example, a change has occurred in the gradient. The shift in the zero point can be illustrated with a linear system in Table 1.1 and Fig. 1.6.

TABLE 1.1 Voltage Determinations Along a Linear Distribution in a Material of Uniform Resistance (Volts) and a Resistance Where the Positive Half That Is Twice That of the Negative Half (Volts′)

cm	-10	-9	-8	-7	-6	-5	-4	-3	-2	-1	0	1	2	3	4	5	6	7	8	9	10
Volts	0.50	0.45	0.40	0.35	0.30	0.25	0.20	0.15	0.10	0.05	**0.00**	-0.05	-0.10	-0.15	-0.20	-0.25	-0.30	-0.35	-0.40	-0.45	-0.50
Volts′	0.50	0.47	0.43	0.40	0.37	0.33	0.30	0.27	0.23	0.20	0.17	0.10	**0.03**	**-0.03**	-0.10	-0.17	-0.23	-0.30	-0.37	-0.43	-0.50

Zero point in bold type.

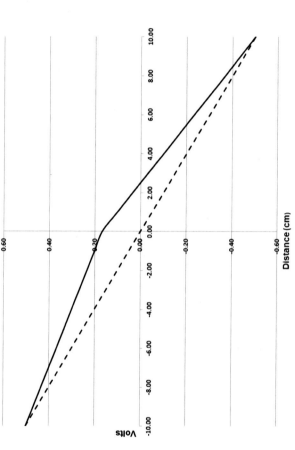

FIGURE 1.6 Graphical representation of the data in Table 1.1. The *dashed line* is taken from the volts data in Table 1.1 and the *solid line* is taken from the Volts' data.

The resistance is the same in each half of the Volts row of data in Table 1.1. This is not true for the Volts' row of data and the lower resistance portion of the example is the left (negative) half from the midposition in the Volts' row of data. The zero point has moved to the right in the Volts' row, the higher impedance half, to 2.5 cm to equalize the resistances on both sides of the zero point (abscissa intersection in Fig. 1.6). This also corresponds with closer spacing of the equipotential lines of force (voltage lines) on the right half seen as the steeper slope of the Volts versus distance data. Since the applied voltage is unchanged, the gradient on the right half of my linear example is 0.5 V/7.5 cm, while on the left side it is 0.5 V/12.5 cm, whereas the gradient was originally 0.5 V/10 cm on each side (the Volts row of data).

It is this increased slope of the voltage gradient, on the right-hand side of Fig. 1.6, that is associated with increased numbers of neurons reaching a transmembrane voltage gradient of 6−7 mV and consequent depolarization by the externally applied field, marked by the increased electric shock sensation in the examiner's finger. Conversely, on the lower impedance side (the subject), the equipotential lines of force have moved farther apart, resulting in a less steep voltage gradient, making nerve depolarization even less likely and explaining why the subject noted no sensation during the procedure. Cooper has provided mathematical support for how externally applied fields accomplish nerve depolarization through the creation of local voltage gradients of sufficient magnitude for action potential generation.[8] Extension of Cooper's analysis also provides the rationale for long observed and inadequately explained strength/duration curves, covered in more detail in Chapter 3.

APPLICATION OF ELECTRIC FIELD THEORY TO NERVE STIMULATION

At this point, a discussion of a situation I have found quite vexing during my years practicing regional anesthesia is appropriate. This relates to how nerves are electrically stimulated in clinical applications such as providing nerve blocks for surgical procedures.

Needle electrode stimulation of nerves is classically depicted something like Fig. 1.7.

In Fig. 1.7, a needle electrode is shown in close approximation to the nerve. The electrode is cathodal and the picture shows negative charge (depolarization) accumulating on the surface of the nerve inducing a neighboring region of positive charge accumulation (hyperpolarization). The only problem with this image is that in ionic media, such as tissue, charge does not just puddle in places. It moves in the form of ions flowing between the electrodes. The current lines, in the medical literature, are frequently confused with electron flow, but ionic media does not conduct current carried on electrons. The only region where electron flow occurs is in the immediate vicinity of the electrode surface, i.e., the first few nanometers. In that region, electrons coursing down

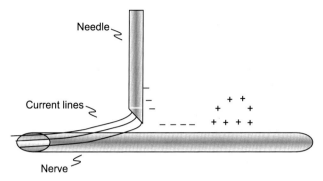

FIGURE 1.7   Classic depiction of nerve stimulation via a stimulating needle electrode.

the metallic needle combine with positively charged ions (cathode) or are released by negatively charged ions and collected by the anode. Electrons do not flow into the media, nor does charge accumulate as shown in Fig. 1.7. Current is not "injected" into the nerve. Charged moieties, ions, that are preexisting both within the nerve cells and external to them simply begin to migrate toward the source of opposite charge in the externally applied field and this migration constitutes current flow. It is the developed voltage gradient, emanating from the needle tip that leads to sodium channel opening, resulting in transmembrane current flow in the form of sodium ions moving from the external side of the cell membrane into the cell interior, and explains the depolarization of the nerve (Fig. 1.8). No charge comes from the needle tip to enter the nerve; the charges all exist in the interstitial fluid surrounding the nerve and intracellularly prior to the needle arriving on the scene.

Note that in Fig. 1.8, along the depicted current path the equipotential lines of force (voltage) are perpendicular to the current lines. This means that at any

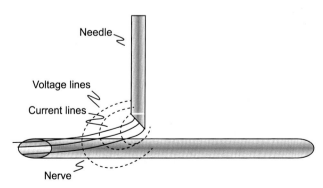

FIGURE 1.8   Situation depicted in Fig. 1.7 with equipotential lines of force (voltage lines) shown.

point along the nerve membrane where the equipotential line is perpendicular to the membrane, e.g., when the current is flowing along the interior of the neuron, the voltage gradient across the membrane is zero and will not lead to depolarization, no matter how much current may be flowing. An example of this will be discussed in a few paragraphs. It is important to remember that as Hille says, the current is irrelevant, though if it exceeds the ampacity of the neuron, it may become quite relevant by heating the nerve.[1]

Close inspection of Fig. 1.8 reveals that the equipotential lines of force are arranged parallel to the membrane in the immediate vicinity of the point of the needle. It is here that a transmembrane voltage gradient of sufficient magnitude to move the charged moieties on the voltage sensor of the sodium channel protein and depolarize the nerve can be established in contrast to where the equipotential lines of force are perpendicular to the nerve.

These observations bring us to the intriguing example of the Griffith needle.[9]

The needle depicted in Fig. 1.9 consists of an outer insulating layer (a) that coats a stainless-steel tube (b), inside of which is a second insulation layer (c) that coats a central electrode (d), constructed so that only the tip, as shown, is exposed. The cathode and anode are the central and tubular electrodes, respectively. It is important to note, at this point, that surfaces in tissue, e.g., a needle surface, are coated with interstitial fluid and that fluid represents a low, ohmic resistance to current flow lacking the impedance barriers of cell membranes. Consequently, current flows preferentially through the low ohmic fluid resistance on surfaces, and this includes the surfaces of internal organs.

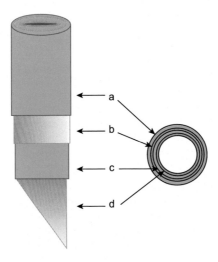

FIGURE 1.9 The Griffith needle consists of an outer insulating layer (a) that coats a stainless-steel tube (b), inside of which is a second insulation layer (c) that coats a central electrode (d).

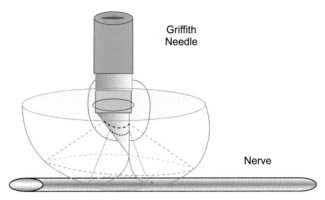

FIGURE 1.10 The Griffith needle shown with current (red) flowing between the cathode and anode from Fig. 1.9 and the torus of developed voltages (green) shown in relation to a nerve.

Current flow in the Griffith needle situation will not go into the nerve at all, but simply courses over the needle surface between the cathode and anode (the red lines in Fig. 1.10 representing cross sections of isocurrent surfaces).[10] The voltage surfaces (green in Fig. 1.10) form a torus around the needle tip, presenting a large region where parallel alignment of the voltage gradient to the membrane occurs. Due to these voltage characteristics, this needle was found by clinical evaluators to be a very effective nerve stimulation device that "injects" no current.

In Fig. 1.10, the 0 V/cm plane is depicted as a disc and two additional equipotential surfaces are shown as truncated cones by the dashed lines. Note that the equipotential surfaces run at right angles to the isocurrent surfaces. The solid green lines represent the voltage torus that surrounds the needle tip. The truncated cones used to show the equipotential surfaces are meant as approximations since the actual shape of those surfaces will be more complex.

The understanding of current flow over the surface of a multipolar electrode array, such as the Griffith needle, is important to applications beyond needle stimulation of nerves. Multielectrode arrays are used in neuromodulation techniques, such as spinal cord stimulation, often shown with current lines spreading far into the surrounding tissue. This simply does not occur. Though a small fraction of the current will flow into the tissue, the vast majority flows over the surface of the electrode assembly, never penetrating the tissue in which the array is imbedded. This is because the current tracks through the low, ohmic resistance of the interstitial fluid coating the array as opposed to traveling through the higher impedance of the surrounding tissue. It's a "path of least resistance" phenomenon.

The situation for multielectrode arrays does not pertain to unipolar electrodes (Figs. 1.7 and 1.8) commonly used for regional anesthesia, or neurophysiologic testing such as nerve conduction velocity studies and electromyography. In those circumstances, the current path must extend across the intervening tissue between the unipolar sampling/source electrode and the remote return electrode. This will be discussed to greater depth in Chapter 3, but a model of tissue as a homogeneous bulk conductor with imbedded electrodes that are either unipolar or multipolar, which lacks the investing layer of interstitial fluid, is misleading and when used as a tacit assumption (as it has for most stimulation applications) it leads to suboptimal system design.

EXAMPLES FROM THE CLINICAL REALM

There exists additional supporting evidence for nerve stimulation being an electric field-mediated event. In a letter to the editor, Vloka and Hadzic bring up two observations regarding current levels and sciatic nerve stimulation.[11] One example is that injection of small volumes of local anesthetic during nerve stimulator operation results in immediate cessation of motor twitch. This is certainly true and my personal experience is that volumes of less than 0.2 mL of solution, local anesthetic or saline, suffice to cause this effect. As Vloka and Hadzic point out, blockade of sodium channels is not an immediate event but requires diffusion time such that an immediate cessation of motor twitch with small volume injection will not occur. Clearly sodium channel blockade by local anesthetic does not explain the finding, especially since it also occurs using saline as an injectate. However, the authors' conjecture that it is mechanical deformation of the tissue, moving the nerve to a greater distance from the needle tip that causes the effect also lacks credibility when such very small volumes of solution suffice for the effect. It's difficult to imagine such small volumes displacing a large structure such as the sciatic nerve. Rather, the ohmic resistance of the injected fluid results in an immediate decrease in the system impedance. Since it is felt necessary that the stimulators used for regional anesthesia be controlled current devices (an example of medical lore rather than scientific fact) with injection, there is also an immediate fall in the voltage required to drive the controlled current and a concomitant decrease in the electric field gradient in the vicinity of the membrane. The fall in the voltage gradient results in inadequate depolarization for action potential generation.

A particularly revealing observation of Vloka and Hadzic is a burning sensation noted at current outputs in excess of 1.0 mA. They credit this to high current density at the needle tip, which seems a reasonable explanation related to thermal effects in the tissue activating wide dynamic range receptors (a class of nerve receptors). However, the interesting thing is that this effect is seen when a stimulating needle is inserted just deep to the epidermis (personal experience). A very high density of axons traveling up to the skin surface exists in this region ($\sim 300,000-350,000$ axons per square inch of skin surface), yet despite these axons being in proximity to the needle tip, stimulation of any sensation, e.g., pain, itch, tapping, cannot be elicited even after trying multiple frequencies and output amplitudes. The reason appears to be the orientation of the electric field equipotential lines of force at approximately right angles to the small axon long axis. Because of this field orientation, there is no transmembrane potential gradient at any one point on the membrane since the potential is equal on both sides. It is not until current levels are high enough that the ampacity of the tissue is exceeded and local heating occurs that any sensation is noted. Once the needle tip is brought into proximity of a deeper fiber running at right angles to the course of the needle (the situation shown in Fig. 1.8), and parallel to the developed equipotential lines of force, that membrane voltage gradients of sufficient magnitude for depolarization and action potential generation are possible.

STIMULATING NEEDLES ARE NOT POINT SOURCES IN SPACE

Impedance also falls with increasing insertion length of stimulating needles and catheters. This is shown in Fig. 1.11 from an experiment where a 24G, insulated needle was inserted, subcutaneously, to 7.5 mm. Impedances were calculated from peak-to-peak voltage differences measured at a sinusoidal controlled current output of 40 μA, with a frequency of 2000 Hz. Though this study demonstrated a somewhat linear relationship between impedance and depth, additional studies with greater insertion depths show that the impedance asymptotically approaches a minimum value with increasing insertion depth.

Clinicians think of stimulating needles as point sources of electrical output, usually charge, though I hope I am providing convincing evidence that charge is not the important factor for stimulating nerves. The idea is that by insulating the needle, the only source of electrical output is the exposed tip of the needle and it can be considered a point source. However, Fig. 1.11 is a good example of why that conception of stimulating needles is incomplete.

The reason insulated needles and catheters are associated with a decreasing impedance with increasing insertion depth is that they are variable capacitors. The structure involved is a conductor (needle), surrounded by insulation, the whole assembly surrounded by another conductor (tissue). This is a classic capacitive assembly and since impedance has an inverse relationship to capacitance, the greater capacitance associated with greater depth of insertion (and greater length of the capacitive structure) causes a fall in overall system impedance. If these variable capacitance constructs, insulated needles and catheters, are inserted greater than a critical depth, the controlled current output of standard nerve stimulators will not be associated with sufficient driving voltage to create an adequate potential gradient across the neuronal cell membrane to effect depolarization and action potential development. This is a consequence of the voltage to resistance relationship shown in Eq. (1.1).

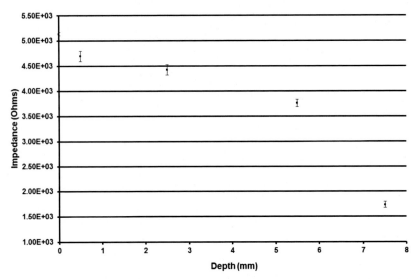

FIGURE 1.11 Calculated impedances from an insulated, 24G stimulating needle inserted in tissue. *(Disclosed in Cory, US Pat. Appl. US20060195158)*

Additional confirmation of the depth relationship was presented by Dr. Hadzic at the Annual Meeting of the American Society of Regional Anesthesia in 2006. During his presentation, Dr. Hadzic noted as an aside that greater current outputs were required to stimulate deeper lying nerves for unknown reasons. From the foregoing discussion of variable capacitive effects related to insulated stimulating needles, this is an obvious consequence since more current output will be needed to create adequate field strengths in the vicinity of the more deeply placed nerves.

Another confirmation of the capacitive effects is the loss of stimulation from stimulating catheters while being threaded to greater lengths. This is well known and assumed to be caused by catheter tip migration away from the nerve as threading progresses. However, such migration is frequently not seen when using simultaneous ultrasound visualization of the catheter during the threading process. The explanation is again related to the variable capacitance of the long, insulated stimulating catheters and decreases of driving voltages with greater insertion lengths. It is unfortunate that lore has developed in clinical nerve stimulation regarding the necessity of constant current outputs, though from the discussion of the history of this concept, the reason for it is apparent. It is a result of thinking that current or charge is the important factor in neuronal cell membrane depolarization and that constant current outputs are necessary to compensate for impedance changes. This is actually just the opposite of what will be most effective: controlled voltage outputs to create consistent field strength near the tip of the stimulating construct (needle or catheter).

There is yet another consideration in all of this. When a 22G stimulating needle is used as an electrode, the tip of the needle has a diameter approximating 0.72 mm depending on the bevel. The inner diameter approximates 0.41 mm, giving an area of the ring-shaped metallic electrode of around 0.28 mm^2. Geddes work demonstrates that for stainless-steel electrodes, when the current density exceeds 10 mA/cm^2 both the capacitance and resistance show marked nonlinear behavior with the capacitance steeply increasing with increasing current density and the resistance falling just a steeply.[12] The net result is that the impedance of the electrode-tissue system will fall dramatically as the current density increases beyond 10 mA/cm^2. So, at the usual 0.5 mA output of a stimulating needle, thought necessary for adequate nerve stimulation, what is the current density? It's 182 mA/cm^2! Clearly, the voltage developed with the use of a controlled current nerve stimulator driving a current density of 182 mA/cm^2 will be extremely nonlinear and change precipitously with very slight changes in current levels. Now, consider what happens if the needle, rather than being empty (air-filled) is instead flushed with electrolyte (local anesthetic solution or saline). Suddenly, the participating surface area of the electrode is increased by the internal surface area of the needle that is in contact with the electrolyte. For a typical 2″ long, 22G stimulating needle, that would amount to adding 0.65 cm^2 of surface, giving a total of 0.6528 cm^2 and a current density at a 0.5 mA output of 0.77 mA/cm^2, well within the linear range of stainless-steel electrodes.

The point of the above discussion is that baseline conditions affect the developed voltage from a stimulating needle in a marked fashion. The take-home message is that whenever one performs an injection procedure using nerve stimulation for needle tip placement, the needle should be flushed with local anesthetic or saline prior to initiating stimulation.

The recognition of the importance of the orientation of the electric field to the neuronal cell membrane also explains the observation that larger anodal currents are required to stimulate nerves than those delivered at the cathode. This is most often rationalized as a function of charge distribution on the membrane similar to that depicted in Fig. 1.7. The proposed mechanism is related to positive charge accumulating on the membrane closest to the anodal source electrode, which then induces a region of neighboring negative charge that causes depolarization. Actually, the voltage gradient in the region of the needle tip that causes sodium channel opening is also responsible for any charge redistribution, not vice versa, as discussed extensively above. The electric field causing the transmembrane voltage gradient across the near membrane also causes a voltage gradient profile across the whole of the circumferential membrane. The gradient will decline from its maximum next to the needle source to a value of zero at 90 degrees and 270 degrees, then increase to a maximum of opposite polarity at 180 degrees. Since the magnitude of the gradient will be less on the opposite side than the side closest to the needle simply as a reflection of the distance involved, a larger current output is required from an anodal source to cause a gradient of sufficient magnitude on the other side of the neuron to effect depolarization and cause action potential generation.

What becomes clear from the foregoing is that the provision of a constant flow of current, or charge, is not the relevant factor in stimulating nerves, recognized by neurophysiologists such as Bertil Hille and many others for more than 60 years. It is, rather, the ability to generate an electric field of sufficient magnitude across the neuronal cell membrane that results in sodium channel opening and drives action potential generation. To think that the only way to accomplish this is to line up charges of opposing values on opposite sides of a membrane, though consistent with the potassium ion observations of over 100 years ago, is to not understand the ways in which electric fields may be generated. It is essential to realize that an electric field for nerve depolarization may easily exist with no charge movement, or current flow, whatsoever, e.g., in an oil bath, and yet cause depolarization and action potential generation in nerves suspended in the oil. Observations from the Griffith needle, depth of stimulating needle insertion, injections of small volumes, and the TENS technique all lend support to the hypothesis that nerve stimulation is a field effect, unrelated to the current flow, point charge accumulation, or the absolute value of the voltage applied—and not at all the subject of dragons or mystery.

REFERENCES

1. Hille B. Ionic basis of resting and action potentials. In: Kandel ER, editor. *Handbook of physiology; Section 1: the nervous system*, vol. 1. Bethesda, Maryland: American Physiological Society; 1977.
2. Kaslow A, Lowenschuss O. Dragon chasing: a new technique for acupuncture point finding and stimulation. *Am J Acupuncture* April−June 1975;**3**(2).
3. Kwok G, Cohen M, Cosic L. Mapping acupuncture points using multi channel device. *Australas Phys Eng Sci Med* 1996;**21**(2).
4. Cory P, Bigeleisen P. Impedance neurography. In: Bigeleisen P, editor. *Ultrasound-guided regional anesthesia and pain medicine*. Philadelphia: Wolters Kluwer|Lippincott Williams & Wilkins; 2010.
5. Reichmanis MMABR. Electrical correlates of acupuncture points. *IEEE Trans Bio Med Eng* 1975;**22**(6):533−5.
6. Rosell J, Colominas J, Riu P, Pallas-Areny R, Webster J. Skin impedance from 1 Hz to 1 MHz. *IEEE Trans Biomed Eng* August 1988;**35**(8):649−51.
7. van Nostrand R, Cook K. *Interpretation of resistivity data*. Washington, D.C.: United States Government Printing Office; 1966.
8. Cooper M. Membrane potential perturbations induced in tissue cells by pulsed electric fields. *Bioelectromagnetics* 1995;**16**:255−62.
9. Griffith R, Strowe R, Newell J, Edie P, Messina R, Houghton F. *Inventors bi-level charge pulse apparatus facilitate nerve location during peripher nerve block procedures*. December 29, 1998. 5853373.

10. Rudy Y, Plonsey R. The eccentric spheres model as the basis for a study of the role of geometry and inhomogeneities in electrocardiography. *IEEE Trans Biomed Eng* July 1979;**26**:392—9.
11. Vloka J, Hadzic A. The intensity of the current at which sciatic nerve stimulation is achieved is more important factor in determining the quality of nerve block than the type of motor response obtained. *Anesthesiology* 1998:1408—10.
12. Geddes LA, Da Costa CP, Wise G. The impedance of stainless-steel electrodes. *Mol Biol Eng* 1971:511—21.

Chapter 2

Skin Surface Impedance

Among the lessons learned during the development of Impedance Neurography was that subtleties, initially thought to be of little importance, had substantial potential to affect measurements. The construction of the system electronics and the analysis software were compelling, intellectual challenges that easily preoccupied the development team, but we found seemingly more mundane issues involving the nature of the contact between the measurement system and the skin surface as well as the surface area of that contact, substantially affected the outcome. Moreover, the literature was confusing regarding these issues and did not provide a great deal of guidance. In the following sections I will discuss some of these "mundane" issues that turned out to be quite critical in the end.

THE NATURE OF SKIN CONTACT SYSTEMS

Skin impedance determinations are subject to variable conditions that affect the final observation. There are many considerations for adequate electrical contact regarding the skin including the nature of the interface between the skin and the electrode as well as the electrode composition itself, but the most important thing for comparing site-to-site impedance is that the measurement method must be consistent with recognition of its limitations. A critical finding for Impedance Neurography was the discovery that the interface between the electrode material and the skin surface must be aqueous, though it took some time to figure this out. An understanding of the physical chemistry involved was essential for this understanding since Impedance Neurography represents an intersection between physics, physiology, and chemistry.

The Transcutaneous Electrical Nerve Stimulation (TENS) technique demonstrated that coupling the electrode (the examiner's finger) to the subject via conductive gel was effective for making observations relative to the position of underlying nerves and nerve-related structures. But, as I mentioned in Chapter 1, some clinicians had no ability to detect the sensation changes in their fingers, besides which this was a messy proposition requiring paper towels for postexamination clean up as well as the gel being a cold experience for the subject. Part of the development of more sophisticated instrumentation than a finger began with inquiring into alternative methods of contacting the skin surface.

Finding the Nerve. http://dx.doi.org/10.1016/B978-0-12-814176-2.00002-2

There are devices on the market called "trigger point locators," and early studies with those devices that used bare metal electrodes were not successful in identifying nerves. Additionally, the use of those devices was often associated with very noticeable pinprick sensations. Delving into the literature regarding electrode systems, it became apparent that there are two, well-described reasons for the ineffectual nature of those systems: the rapid development of thin oxidation films on the electrode surface due to skin surface moisture content and the occurrence of high, regional charge densities on the metal surfaces.

Oxidation films on the metal surface act through the half-cell potential of the particular metal comprising the probe to create a back emf opposite to that of the applied current, effectively degrading the developed voltage and creating a source of error since the degree of back emf formation is neither controllable nor predictable. The issue of regional charge density anomalies that occur in metal electrodes is due to metallic composition variation and resultant nonhomogeneities in conduction characteristics of the electrode. Those charge density differences resulted in most the current flowing from a small portion of the electrode surface, hence the pinprick sensations.

Complicating the problems associated with bare metal electrodes was a determination that skin surface impedance involved a transition between the metallic conduction (electron flow) of the measuring system and the ionic conduction of the tissue electrolyte. An aqueous gel interface clearly accomplished this transition, albeit with a significant fiddle factor as our system used a wet gel that had to be pipetted into wells in an acrylic, closed cell foam skin attachment system. This attachment system adhered to the electrode array surface on one side and the skin on the other. Also, care had to be taken to ensure that no gel bridged across the foam matrix from one well to a neighboring well, which would effectively double the skin contact area for both electrodes. Preventing bridging was accomplished by having a polyethylcne release layer on the skin contact side of the foam matrix. After filling all the wells with gel, the excess was squeegeed off so the gel was flush with the release layer surface. Then the release layer was carefully peeled off to leave each gel surface two thousandths of an inch proud of the foam matrix surface. Care also had to be observed so that small bubbles of air were not left in contact with the electrode surface that decreased the contact area and changed the impedance determinations. Doing all of this worked, but for a while the admonition, "the enemy of good is better," was ignored and the search was on for a better technique.

This process led to the discovery that z-axis—specific, conductive polymer sheets were available. This is some very clever technology wherein silver slivers are embedded in a carbon-doped silastic sheet so that the long axis of the silver slivers is aligned with the thickness of the sheet: the z-axis. The difference between the conductance of the silver slivers versus that of the carbon-doped silastic allows preferential current flow across the thickness of the sheet. When placed on our electrode array and attached to the skin surface the material was found to be entirely ineffectual in allowing skin surface impedance determinations. At the time, though the reason for the lack of effectiveness was not apparent, we continued our search for alternatives.

About this time, we noted the existence of a medical device called the T-Scan 2000 ED. This device was marketed for the diagnosis of cancer via skin surface impedance determinations first by Siemens, and then by Mirabel Medical. The electrode system used consisted of an array of square electrodes in a handheld assembly that was filled with conductive gel and then slid back and forth over the skin surface to purportedly detect regional impedance variations reflective of underlying malignant processes using a two-electrode system that was similar to ours. Though doubtful regarding how such a skin interface system could work since the electrical contact area for each electrode was essentially uncontrolled, the company reported significant results and it seemed worth a try.

For our purposes, some polypropylene mesh, such as used for tissue defect repairs in hernia operations, was attached to the electrode array surface by first spraying glue on the mesh, then pressing it on the electrode array. Conductive gel was smeared over the mesh and squeegeed in. Since the mesh strands were very much smaller than the electrodes, the amount of electrode surface covered by mesh averaged out. Then the electrode-mesh system was carefully applied to the skin surface so as to prevent any lateral movement during application and taped securely in place on the subject. Comparing measurement results from this system versus the individually filled wells of our original system demonstrated that the mesh system allowed too much bridging between electrodes to be of any use. This also probably played a role in why the T-Scan 2000 ED was never successful, though an even greater role involved the electrical characteristics of the tissues involved, which will be discussed in much more detail in Chapters 3 and 4.

In the final analysis, what it came down to was that to effect a reproducible electrical transition, an aqueous interface that wetted the surfaces involved and at the same time controlled skin surface contact area was an absolute requirement. This physicochemical process of wetting allowed an efficient transition from electron flow through the electrodes to ionic conductance in the tissue with the actual transition occurring in the gel material. Using this approach, by the time the current flow reached the tissue, it was already ionic in nature. However, a second process involving the skin surface also occurs during electrode contact with an aqueous interface; that same aqueous medium also hydrates the skin resulting in a progressive reduction in impedance determinations with time as the skin imbibes water. Fortunately, the physicochemical process of surface wetting occurred rapidly following exposure to the aqueous medium, while hydration was not complete even after 6 hours of continuous application. The different time courses for the two effects meant that within a few seconds of electrode application, accurate impedance measurements were determined without significant error introduced by the much slower process of hydration.

SKIN SURFACE CONTACT AREA

When first making observations using the TENS technique with a finger as a probe, a light touch was found important since too great a pressure tended to cause sensations that masked some of the changes in electric shock perception. When a "light touch" was evaluated it corresponded to a contact area of about 0.5 cm^2 so that size was used as the early electrode contact area. This turned out to be a fortuitous, though entirely uninformed decision. Studies prior to that time reported skin impedance values to be constant wherever they were determined on the body.[1] These studies employed standard electrocardiographic electrodes with contact areas in excess of 1 cm^2. I found that there was at this time (1980s) an instrument called the Sentri that used a much smaller contact for locating myofascial trigger points. This was one of several so-called trigger point locators, but the Sentri used a moistened felt interface with a diameter of about 4 mm between a stainless-steel electrode and the skin surface. Experimentation with this device revealed that, if one was very patient, it could be used to determine the location of nerves in contrast to the studies with devices using the larger diameter ECG electrodes. Further, the trigger point locators employing bare metal electrodes were essentially useless because of the factors listed above as well as the variable nature of metal electrode capacitance and resistance that are both contact area and current density dependent.[2,3] For metallic electrodes, both capacitance and resistance are linear with slopes near zero for a range of contact areas and current densities, specific for the composition of the electrode, e.g., stainless-steel versus gold versus Ag:AgCl. However, outside those ranges, both electrical parameters become nonlinear, a fact that places constraints on the lower size of effective electrode surface areas.

MEASUREMENT TECHNIQUES OF IMPEDANCE TOMOGRAPHY AND IMPEDANCE NEUROGRAPHY

While discussing electrode contact area, it's important to keep in mind that skin impedance determination assesses the impedance of all the tissue components in the electrical path between the skin surface electrodes in a two-electrode setup such as used in Impedance Neurography. This two-electrode approach is different from the technique employed for Impedance Tomography, which can be confusing, especially for investigators working with the latter technique who have become used to a three-electrode system. Some explanation is in order.

Impedance Tomography is an approach to imaging structures beneath the skin surface by passing current between two electrodes, usually located on the foot and the head of the subject, then measuring the developed voltages occurring on the skin surface at points between the current-carrying electrodes. This can be illustrated with Fig. 2.1, which was discussed in Chapter 1.

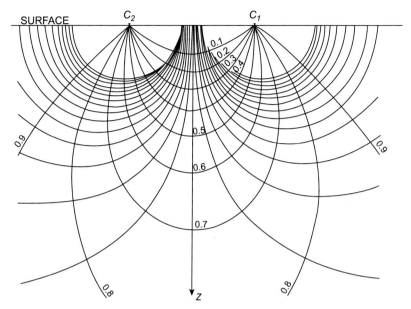

FIGURE 2.1   Classic electric field distribution in a bulk conductor.[4]

In Fig. 2.1, a voltage difference is established between the two electrodes, C_1 and C_2. This creates an electric field in the intervening material through which current flows, indicated by the isocurrent lines connecting the two electrodes. The electric field in V/unit distance results in discrete voltages developing at the surface on which the electrodes are placed. The voltage differences between any two points (third electrode) on the surface can then be measured, and depending on how those voltages are distributed, through the use of complex mathematical analyses called "back projection algorithms" assessments of the impedance characteristics of the underlying material can be made. In Fig. 2.1 the material is homogeneous, and the resulting surface voltage distributions shown between C_1 and C_2 are uniform. If there were discrete regions of differing impedance in the material of Fig. 2.1, the surface voltage distributions would not be uniform but would be skewed as a reflection of how the current flows through regions of variable impedance.

The technique of Impedance Neurography determines the impedance between the electrodes C_1 and C_2 and is not concerned with the surface voltage distribution between those electrodes. It is apparent that the use of this approach means that measurement of the impedance of the skin, in isolation, is not possible in an intact person by applying two electrodes to the skin surface and passing current between the electrodes because the current path includes the subcutaneous structures in addition to the skin. As a consequence of the way the electric field distributes, more tissue than just the skin participates in impedance determinations. This approach of using two-electrode impedance determinations is one of the flaws in some galvanic skin resistance or GSR measurement devices, which do not take into account the subcutaneous contribution to the impedance of the total electrical path. It so happens that current does not quite distribute in tissue the way Fig. 2.1 depicts, but for right now it's a useful representation indicating the large contribution of subcutaneous structures in the surface impedance determinations. However, those differences in current distribution are what allow nerve-specific information to be extracted from such an approach. This will be clarified in Chapters 3 and 4.

There are a couple of considerations regarding contact area of the electrode on the skin surface that may not be obvious at first pass: contact impedance and skin surface impedance integration.

CONTACT IMPEDANCE

Contact impedance results from the nature of the contact between the electrode and the surface and affects the measured impedance. This relationship between contact and impedance is not unique to interactions with the skin. There are multiple factors contributing to contact impedance that can make measurement of very small signals quite complex. Fortunately, the signal-to-noise ratio in Impedance Neurography is large, in the range of 0.5–2 orders of magnitude, so many of the complicated capacitive components of contact impedance are not significant sources of error. There is one factor that does place constraints on Impedance Neurography measurement systems: the area of contact. Fig. 2.2 shows the calculated impedance between two electrocardiographic electrodes

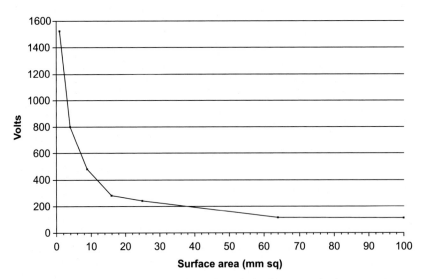

FIGURE 2.2 ECG electrode surface area versus applied voltage to maintain constant current flow. *(Disclosed by Cory in US Pat. Appl. #20120323134)*

with a series of polyethylene masks interposed between the electrodes. The masks had holes of varying diameter allowing contact of the two hydrogel components of the electrode assemblies. Importantly, the Ag:AgCl electrode—electrolyte interface contact area remained constant in this experiment.

The method employed for measurement in Fig. 2.2 consisted of applying a constant current source and measuring the applied voltage required to maintain the constant current. Since the applied voltage to maintain constant current varies directly with the system impedance, increasing voltage indicates increasing impedance. What is obvious from Fig. 2.2 is that the applied voltage versus contact area is a power function and that the curve becomes quite steep for field path cross-sectional areas of less than 10 mm^2. If contact areas are reduced below that level, the measurement of the applied voltage becomes critical as small deviations in surface contact between electrodes can result in large errors in calculated impedance. As electrode contact area is decreased, the contact impedance begins to approach the values of the total system impedance. Thus, there exists a practical lower limit of about 3 mm for the electrode diameter below which measurement error becomes significant.

SKIN SURFACE IMPEDANCE INTEGRATION

Skin surface impedance integration refers to what I call the unit skin surface impedance area. Our investigations revealed from the first time the TENS technique was used that some electrical characteristic of the skin varied from point to point and it turned out that factor was impedance. What was determined was the impedance of the skin surface is a mosaic structure that is similar to the diagrammatic representation in Fig. 2.3.

The group of red and blue hexagons in the middle of Fig. 2.3 shows an equal distribution within the larger, black hexagon. If the red hexagons represent high unit impedance sites, while the blue are low unit impedance sites, the net impedance calculated from within the larger hexagon will be an integrated value that averages the highs and lows. It is obvious that as the measurement area is reduced, and with appropriate positioning of the reduced hexagon, a preponderance of high or low small hexagons could be included within its boundaries and the integrated value would be higher or lower depending on just where the measurements were obtained.

At the extreme, the measurement hexagon size could be reduced to the point that it is wholly contained within the boundaries of a small high or low unit impedance hexagon. It is this integration effect that partly accounts for earlier observations that skin impedance was the same no matter where on the skin surface it was measured with relatively large contact area electrodes. Now, consider a situation more akin to the observations obtained using the TENS technique over a nerve, depicted in Fig. 2.4.

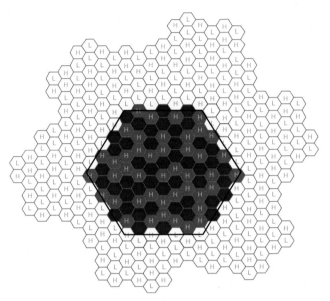

FIGURE 2.3   Representation of skin surface impedance distribution.

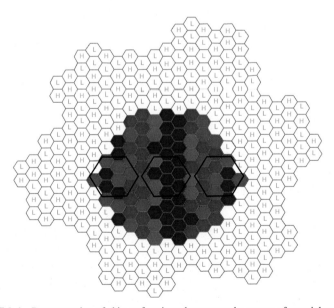

FIGURE 2.4   Representation of skin surface impedance over the course of a peripheral nerve.

In Fig. 2.4, three small hexagonal regions, showing three different positions of a fingertip, are represented on the skin surface overlying a peripheral nerve coursing at right angles to the line of the three, small hexagons. Rather than just two varieties of unit impedance areas (high and low) a more representative situation is depicted where intermediate impedance value sites (shades of red) are interposed between the line of low value sites overlying the nerve and higher sites on each side of the nerve. It is clear that the impedance value derived from the center hexagon will be lower than that of either edge hexagon. The actual size of a unit impedance skin surface area is not known at this time primarily due to the considerations associated with skin contact area impedance placing practical limits on the size of measuring electrodes.

A real-life example of this effect is shown in Fig. 2.5, an impedance neurograph obtained from the posterior, distal thigh about 10 cm proximal to

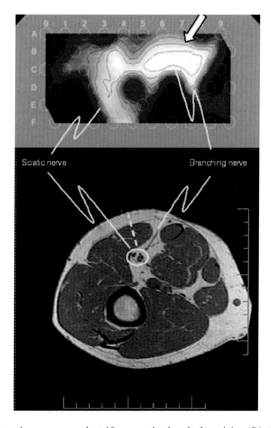

FIGURE 2.5   Impedance neurograph at 10 cm proximal to the knee joint. *(Disclosed by Cory in US Pat. Appl. #20060085048)*

the knee joint. The sciatic nerve, clearly showing the peroneal and tibial portions, is contained in the yellow circle on the MRI image. The MRI image slice is a view across the prone limb, e.g., a transverse section through the distal thigh.

The position of an Impedance Neurography electrode array can be discerned as a row of bright dots on the skin surface to which the dashed, yellow line is directed. The reason the electrode positions are revealed as bright dots is that an aqueous gel interface material connected the electrodes to the skin surface and the water contained in the gel was seen as a bright signal on a T1-weighted MRI image, such as in Fig. 2.5. Note that the dashed, yellow line demonstrates the right-angle relationship as it extends to the left-mid portion of the electrode array.

The impedance neurograph shown in Fig. 2.5 represents a longitudinal view of the portion of the thigh shown in transverse section on the MRI image. In other words, the plane of the impedance neurograph was at right angles to that of the MRI slice. Thus, the sciatic nerve seen in cross section on the MRI image appears as a linear structure in the neurograph. Here is where the impedance neurograph gets interesting…

Note the linear structure in the MRI as well as the neurograph to which the label "branching nerve" is directed. When I first saw this MRI image, I asked the radiologist reviewing the image what that linear structure running away from the sciatic nerve might be. He replied that he was not certain, but it was likely a vessel. This is where Impedance Neurography can clarify imaging studies because the neurograph clearly shows a nerve running in the same region where the MRI shows a linear structure. Since Impedance Neurography has no ability to image blood vessels, and the linear structure is a very bright feature of the neurograph, it is clearly a nerve.

A FINAL OBSERVATION FOR CONSIDERATION

The impedance neurograph in Fig. 2.5 demonstrates that a gradient of impedance values was seen to either side of the lowest values (white arrow) just as shown diagrammatically in Fig. 2.4. The presence of such a gradient associated with lateral distance from the position directly overlying the nerve is a clue that there is a relationship between not only lateral distance but also the depth of the nerve beneath the skin surface. The underlying mechanism for the impedance gradient seen to either side of the exact position of the underlying nerve will be discussed in other chapters along with why discrete clumps of low impedance cells, e.g., cancer cells, will not be demonstrated with skin surface impedance determinations, whereas low impedance nerves (linear structures) will be shown, and why the T-Scan 2000 ED was doomed to failure.

REFERENCES

1. Rosell J, Colominas J, Riu P, Pallas-Areny R, Webster J. Skin impedance from 1 Hz to 1 MHz. *IEEE Trans Biomed Eng* August 1988;**35**(8):649—51.
2. Geddes L. Historical evolution of circuit models for the electrode-electrolyte interface. *Ann Biomed Eng* 1997;**25**:1—14.
3. Cory PC. The physics of nerve stimulations and nerve impedance. In: Bigeleisen PE, editor. *Ultasound-guided regional anaesthesia and pain medicine*. 2nd ed. Philadelphia, Baltimore, New York, London, Buenos Aires, Hong Kong, Sydney, Tokyo: Wolters Kluwer; 2015.
4. van Nostrand R, Cook K. *Interpretation of resistivity data*. Washington, D.C.: United States Government Printing Office; 1966.

Chapter 3

The Varieties of Neuronal Cell Membrane Reactance: Nerves as RLC Circuits

So far, the discussion has dealt with practical matters of how electrical fields distribute in tissue and the factors that affect getting current in and out of tissue for reproducible measurements. Now it's time to delve into the underlying electrophysiologic properties of neurons and their cell membranes enabling Impedance Neurography. These investigations started with my assumption that the classic descriptions of tissue equivalent electrical circuits were accurate, but discordant data revealed an entirely new perspective regarding the neuronal membrane equivalent circuit. This section also is pretty technical, but worth plowing through because of the implications of the discoveries. The work described herein has enabled an understanding of how high frequency stimulation techniques for pain control may be improved, pointed a way to improving the algorithms used in Impedance Tomography, provides a rationale for the heretofore poorly explained strength—duration relationships for nerve stimulation, and shows a means of extracting nerve-specific time constant information from tissue.

ELECTRICAL MODELING OF NEURONAL CELL MEMBRANES

The neuronal cell membrane, like all biological membranes, displays the electrical characteristics of resistance (R) and capacitance (C). This is shown in Fig. 3.1 depicting a neuronal cell membrane as a parallel RC (p-RC) circuit.

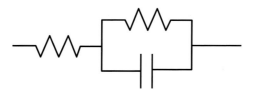

FIGURE 3.1   Parallel RC circuit representation of a neuronal cell membrane.

Finding the Nerve. http://dx.doi.org/10.1016/B978-0-12-814176-2.00003-4

Many studies over the years have shown the neuronal cell membrane to conform to this equivalent circuit, well discussed by Cole.[1] However, a previously *unrecognized* feature of the neuronal membrane equivalent circuit model is its dependence on the characteristics of the applied electrical field. This is the major finding of Impedance Neurography.

Since I have mentioned him, just exactly who was Kenneth Cole? In his 1984 obituary in the *New York Times* he was described as the "father of biophysics." In no small part, we owe much of our understanding of how nerve cell membranes are excited and conduct information to his work. He began by studying suspensions of simple structures such as oil droplets and red blood cells, which led to sophisticated studies of the squid giant axon, though his studies were interrupted for a time during WWII when he worked on the Manhattan Project. His work was critical to our understanding of the electrical functioning of the neuronal cell membrane and essential to the "sodium theory" of nerve depolarization.

Prior to Ken Cole, there were developments in Victorian England that played an important role in his researches as well as those of others. It all began with an urge to communicate across the oceans.

A BRIEF HISTORY OF ELECTROTONICS

Remember the transatlantic cable? That was a great achievement of Victorian era science and technology, but at first all was not well in scientific paradise since transmission speeds were abysmal. In 1858, transmission speeds of about 0.1 words per minute were all that could be achieved. What was going on? The problem was recognized by William Thomson (Lord Kelvin) and Oliver Heaviside who determined that imbalances in the capacitive and inductive reactances of the cable caused dispersion of the signal. This is a fascinating story that is well worth looking up. The cable equation resulted from the work of these individuals and the solution to the transmission woes turned out to involve the straightforward expedient of hanging large masses of iron on the ends of the cable.

Oliver Heaviside is an interesting character in history and played a very large role in the development of electromagnetics. He accomplished this almost entirely outside of the mainstream mathematics and physics community of the time. It was fortunate that his uncle, by marriage, was Charles Wheatstone, professor of physics at Kings College London and inventor of the Wheatstone Bridge, among other things. This relationship led Heaviside into telegraphy and also to discovering the work of James Clerk Maxwell into which he dove with a passion. It is because of Heaviside's work that we have the four, vector equations that we know as Maxwell's equations. Heaviside took Maxwell's original 20 equations with 20 variables and came up with the four, vector equations with which we are now familiar. In fact, Heaviside, along with J. Willard Gibbs, codeveloped vector analysis. He also patented the coaxial cable and gave us terms such as impedance, conductance, and inductance as well as others. The Heaviside Layer, referred to in the musical *Cats*, is a conducting layer in the atmosphere (part of the ionosphere), posited by Heaviside in 1902 and demonstrated in 1923.

Despite his very useful electrical insights that derived from an initial interest in improving the way telegraphy worked, Heaviside did not have a very easy life, a situation that was not helped by his odd behavior and reported inattention to personal hygiene. In that regard, he was not so different from other great innovators such as Beethoven. So why is his work so important to nerve stimulation? Answering that question begins with looking at the cable equation itself.

One of my observations in reading scientific literature is if an article or book has complex equations on the first few pages, it's unlikely that I am going to have an easy time reading the work. In fact, it often seems that the use of such equations, and the accompanying impenetrable prose, often serves to obfuscate more than to clarify. But even though such equations can be off-putting to those of us less mathematically inclined, sometimes it is possible to glean a great deal of information from those mathematical expressions by inspection without actually having to come up with a numeric solution to them. That is my intention in presenting the cable equation, shown as Eq. (3.1) as adapted for neurophysiology, which is a special case of the more general partial differential equation that Heaviside derived.

$$\lambda^2 \delta^2 V / \delta x^2 - V - \tau \delta V / \delta t = 0 \tag{3.1}$$

What can be learned from examination of the cable equation besides the fact that it contains a bunch of Greek letters? First, V (voltage) is shown to vary with the variation of two independent variables, x (distance) and t (time). Remember that the delta character (δ) simply means a very small change. These variations are well understood by engineers designing electrical transmission cables. The issue with which these engineers are concerned is transmission voltage decrements, and these losses increase with increasing length of the transmission line. Importantly, the developed voltage in a cable does not vary with time if the applied voltage is not varying in time, i.e., the developed voltage will not vary if the applied voltage is a direct current (DC) steady state. For a time-variant current, e.g., alternating current or pulsed-DC current, both variables, space and time, affect the voltage measured at any point along the cable. And what about λ and τ in Eq. (3.1)—what do those symbols represent?

λ is termed the length constant and τ is called the time constant, both of which are dependent on the material from which the cable is constructed. The time constant will be a topic of importance later in this chapter where it is covered in detail. The length constant is the ratio of the electrical area resistivity in Ohms·cm across the sheathing (in the case of a nerve, the neuronal cell membrane) of the cable, r_m, to the sum of the linear resistivity of the cable core, r_i, and that of the external medium, r_e, both measured in Ohms/cm shown in Eq. (3.2). Note that the length constant will have units of cm.

$$\lambda^2 = r_m \Big/ (r_i + r_e) \qquad (3.2)$$

How does one visualize the length constant and what does it describe? Basically, the length constant is a way of looking at the distribution of current inside and outside of the cable. A good analogy, described by a neurophysiology colleague, is that of driving cattle, representing current, down a road bounded on each side by fencing that is in a poor state of repair, which represents the cable. As the cattle are pushed along by the cowhands, some will migrate through the holes in the poorly maintained fence to the area outside where they will encounter additional barriers to travel, e.g., sagebrush. Since travel is more difficult going over and around the brush, some of the migrant cattle will decide to move back onto the road through other holes, while some inside the fencing will continue to migrate to the outside. At any point along the road, the ratio of cattle inside the bounds of the fencing to those outside depends on how difficult or easy it is to both get across the fencing and travel outside or inside. This analogy is descriptive of electrotonic conduction where the cattle represent individual charges that are moving (current) in response to the cowhands pressuring them (voltage) down a road bounded by a fence in poor repair (the neuron with its investing membrane).

Now that I have brought it up, I need to explain electrotonic conduction since it's not widely discussed anymore—thing is, it plays a huge role in nerve stimulation as well as Impedance Neurography. For an in-depth discussion of all of this, see Rall.[2]

Electrotonic conduction of a neuron deals with how the neuron functions like an electric cable. This conduction is not the same as that of the more familiar nerve activity that employs propagated action potentials. The structure of a neuron having a low resistance core, high resistance sheathing, surrounded by low resistance fluid has marked similarities with a typical electric cable and the same mathematics are applicable to both. Part of electrotonics concerns the mechanism by which current gets into the cable and out of the cable, hence the cattle drive example. Electrotonics provides a very important description of the electrical functioning of the neuronal cell membrane.

This was a hot topic in biology for decades. A great account of the investigations can be found in Cole's work, *Membranes, Ions and Impulses*.[1] Particularly helpful was the development of the squid giant axon model in the late 1930s, but these studies were largely sidelined by a seminal event in 1952: the publication of Hodgkin and Huxley's work on the ion fluxes across the neuronal cell membrane.[3] Not only is this, too, a fascinating tale, it is a *tour de force* and scientific cultural literacy; every scientist should read it. Like many things, their work benefited from what came before, e.g., the squid giant axon model, plus WWII developments, in this case high speed electronic components used in radar. Their work turned neuroscience away from electrical studies of the membrane to biochemical studies that eventually led to the discovery of voltage-gated channels and all the descendant discoveries that we take for granted today. The equations that Hodgkin and Huxley derived are still used to describe the functioning of the sodium and potassium channels, and they accomplished this with hand-cranked calculators. It took about 2 weeks to calculate one sodium channel depolarization curve! And, these investigators had absolutely no idea about how the ions were traversing the membrane as channels were unknown at the time. This is scientific discovery at its very best.

Hodgkin and Huxley's work did have an unintended consequence; it moved the focus of neuroscience away from electrotonics, which was not entirely a bad thing given all the great discoveries that have happened since they pointed the way forward, but it allowed a bit of misunderstanding to develop regarding the roles of current and voltage in nerve stimulation among clinicians and engineers.

DISCORDANT DATA EMERGE

Fig. 3.2 shows calculated impedance versus frequency for a p-RC circuit such as depicted in Fig. 3.1.

As predicted by theory, due to the frequency-dependent capacitive reactance, an exponential decay of impedance is seen as frequency (f) increases. The equation governing this relationship is as follows:

$$Z = \left(\frac{1}{R^2} + \frac{1}{X_C^2} \right)^{-0.5} \tag{3.3}$$

where Z is impedance, X_C is the capacitive reactance and is equal to $1/2\pi f\, C$.

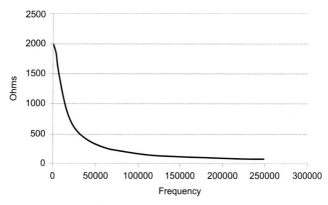

FIGURE 3.2 Calculated impedance versus frequency (Hz) for a parallel RC circuit with a resistance of 2000 Ohms and a capacitance of 10^{-8} F.

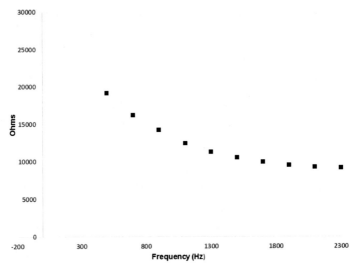

FIGURE 3.3 Impedance (y-axis) versus frequency (x-axis) measured over the saphenous nerve from 500 to 2300 Hz. *(Data courtesy of Nervonix, Inc. Image construction by Philip C. Cory, M.D.)*

Clearly, if frequency is involved in the impedance determination, the electrical field in Volts/unit distance is varying in time, i.e., it is a periodic field. The term, "reactance," indicates a component of the circuit (in this case, the capacitance) is reacting to the periodicity of the field. This reactance occurs because charge accumulates on the capacitance at rate that is dependent on the amount of accumulated charge. As more charge accumulates, the rate of charging slows (due to the repulsion of like charges) and this affects the impedance to current flow across the circuit. With a time-variant field, this reactance affects how quickly charge flows on or off the capacitance and the voltage curve lags the current curve (phase shift). Also, the form of the curve shown in Fig. 3.2 is applicable to situations where the electrodes applying the field are in direct contact with the neuronal cell membrane external surface (external electrodes) or when there is one external electrode and one internal electrode.

Impedance Neurography technology employs an electric field applied at the skin surface, remote from the nerves; it uses an externally applied field. Fig. 3.3 shows a typical impedance versus frequency curve for peripheral nerve derived from this technology.

A HARD LESSON

On initial examination, the curve in Fig. 3.3 looks very similar to that predicted by theory in Fig. 3.2. This similarity gave rise to a long period of confusion, plus a failed licensing agreement for development of Impedance Neurography.

Looking at Fig. 3.3, it appeared that taking measurements of impedance at higher frequencies would be better because I thought that larger separations between the impedances calculated over nerves versus those determined over tissue without significant nerve densities would occur. I began using 2000 Hz as a standard sampling frequency, but obtained confusing results.

Sometimes very good images would result, but this was not reproducible. Though the saphenous nerve on the inner aspect of the calf was well visualized at that frequency, the median nerve in the forearm was not seen at all. This was strange because an earlier device that had a fixed frequency of 250 Hz demonstrated the median nerve without difficulty. Unfortunately, I was quite busy with a clinical anesthesia practice and did not spend time reflecting on those observations. That combination of circumstances led to an understanding of the difference between engineers and biologists since we were working with an engineering group on a licensed development project.

Engineers have the task of taking facts of science and creating new and useful products using those facts. To make certain that the building designed by the engineers would not fall down, or the device will perform as advertised, those facts must be unambiguous and dependable. As a consequence of those demands on engineers, whether mechanical, electrical, or civil, they abhor ambiguity.

Biomedical scientists and physicians, on the other hand, if not exactly embracing ambiguity, at least recognize that it is a fact of life. And for scientists, in particular, ambiguity is great because it is in ambiguous results that true discovery often lurks.

So, though I was quite intrigued by the inability to reproducibly image nerves at 2000 Hz, the engineers were less than thrilled about the observation. The net result was that the licensing agreement fell apart with some significant consequences for us. It was a difficult business lesson, but it led to the most important discovery regarding Impedance Neurography.

In the aftermath of the failed business deal, I began investigating why the median nerve was proving so problematic to image. After finally reviewing observations regarding frequency, I tried to image the median nerve at 500 Hz. A reproducible, beautiful image resulted. What was going on? All the theory about p-RC circuits, which was the accepted electrical circuit model of neuronal cell membranes, seemed to predict that the median nerve should show up just as well at 2000 Hz as it did at 500 Hz.

However, a difference emerged when sampling over a range of frequencies. With a greater extent of frequency sampling, the graph in Fig. 3.4 was obtained. A cursory examination of Fig. 3.4 shows that the impedance did not smoothly approach a lower asymptote as shown in Fig. 3.2. Also, there was a peak of impedance that occurred around 3000 Hz, which is not representative of a p-RC circuit but demonstrates additional parallel resonance phenomena, which require some explanation.

Electrical resonance occurs when two opposing types of reactance exist in a circuit, classically called capacitive reactance and inductive reactance. At the

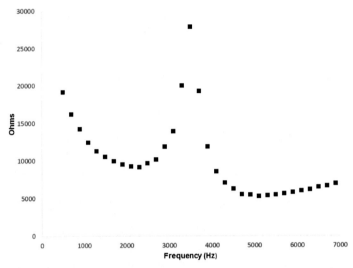

FIGURE 3.4 Impedance versus frequency measured over the saphenous nerve from 500 to 7000 Hz. *(Data courtesy of Nervonix, Inc. Image construction by Philip C. Cory, M.D.)*

resonant or critical frequency, the two reactances are equal in magnitude, resonance occurs, and the circuit electrically oscillates. If the reactances, for an ideal circuit, are arranged in parallel the impedance at the resonant frequency is infinity, while if the reactances are arranged in series the resonant frequency impedance is zero. This property, resonance, turns out to be very useful in circuit design for filtering applications to select specific frequencies or to reject noise.

The form of the curve in Fig. 3.4 clearly represents parallel resonance, but there is an interesting problem in a biologic system such as a neuronal cell membrane, there is no source of Maxwellian magnetic inductance. What accounts for the resonance phenomenon? Also, during the ill-fated licensing agreement, frequency studies had been performed with a commercial gain-phase analyzer and no such resonance was observed. Once again, much head scratching was involved to determine what was being done differently in our system. It turned out that we made a critical design decision, very early on, when trying to mimic the output of the Transcutaneous Electrical Nerve Stimulation unit.

MYSTERY SOLVED: THE EUREKA OBSERVATION

A critical aspect for the observed resonance effect is that the waveform must have a DC offset, and this DC offset must be of a particular form, not heretofore recognized. I will call this a "charged-DC or c-DC" offset, which will be explained later. Though different periodic waveforms (sinusoidal, rectangular, or saw tooth) may be used, these superimposed time-variant fields must exist in combination with the c-DC offset for the resonance effect to occur. The resonance phenomenon will not be observed if impedance versus frequency scans are performed with commercially available gain-phase analyzers. The reason the c-DC offset is important is that another source of reactance is present in the membrane: the ionic conductances.

Resistance and impedance have been mentioned previously, but what is conductance? Well, it's just what one would expect, the ability to conduct charges and that sounds like the opposite of resistance to the movement of charge. In fact, conductance (G) is mathematically the reciprocal of resistance so that $G = 1/R$. Impedance has an analogous reciprocal that is called admittance (Y), and $Y = 1/Z$. A discussion of the conductance of a voltage-gated channel in the neuronal cell membrane relates to its ability to allow charges to pass through the channel and its reciprocal (R) relates to the resistance to charges passing through the channel.

Though we are accustomed to thinking of reactance as capacitive or inductive, in fact any electrical element varying its electrical resistance in response to current is reactive. For instance, an electric filament in a light bulb increases resistance as it heats in response to current flow (thermopositive); it is reactive. Likewise, a thermistor varies resistance with current, though opposite of the electric filament (thermonegative), and is also reactive. Both elements display current-dependent, time-variant resistances.

Fig. 3.5 shows a schematic representation of a neuronal cell membrane with imbedded ionic conductances where C_M is the membrane capacitance, G_K is the potassium conductance, G_{Na} is the sodium conductance, and G_L is the leakage conductance, all represented by a battery in series with a resistance, but displaying different polarities (G_{Na} and G_K). Note in Fig. 3.5 that only G_K and G_{Na} have time-variant resistances.

It's important to understand that Fig. 3.5 is not a representation of individual channels, but rather the overall conductance of the membrane. Consequently, populations of each channel variety may be open, closed, or inactive at any particular moment in time, i.e., the model employs lumped

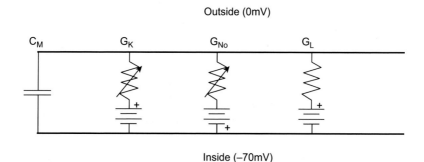

FIGURE 3.5 Schematic representation of a neuronal cell membrane.

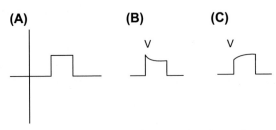

FIGURE 3.6 Input transmembrane current pulse (A) and resultant voltage responses of ionic conductances (B and C).

conductances. Since each battery exerts a fixed potential, there is current flow (ion movement) in both directions through the channels across the membrane, consequently microcurrents do exist, but since the same number of like charges are moving in both directions across the membrane, assisted by the Na−K-ATPase, the net nondepolarized transmembrane current flow is zero.

Fig. 3.6A shows an input, transmembrane step current pulse and associated voltage curves for the two, oppositely directed ionic conductances. If the input current flow is opposite to that of an ionic conductance, due to the current-dependent, variable resistance of the channel, progressively less current will flow through the conductance during the time of the current pulse, decreasing with the increasing magnitude of the time-variant resistance during the current pulse.

Since the conductance battery is exerting a fixed emf, the total emf is reduced by the oppositely directed emf of the externally applied field directed across the variable resistance (Fig. 3.6B). If, however, the input current flow is in the same direction as the ionic conductance, the developed voltage across the ionic conductance increases with the time-variant resistance change during the pulse as shown in Fig. 3.6C. The configuration in Fig. 3.6B is an inductive-like reactance, while Fig. 3.6C shows a capacitive-like reactance; these time-variant conductances can be mathematically modeled as inductive and capacitive re actances. Mauro discusses this in much greater detail and important considerations from his 1961 paper, for the purposes of explaining Impedance Neurography, are included in italics.[4]

> *To display reactance, an element must permit either current or voltage to assume a zero value without both being zero identically. This is a necessary and sufficient condition for reactance. However, a purely dissipative element is characterized by the absence of a mechanism for the storage of electrical energy such that during some part of the cycle energy is absorbed from the source and during another fed back to the source. Thus, in a purely dissipative element the voltage and current must attain zero identically and, therefore, a pure resistance cannot introduce a phase shift between the voltage and current at the fundamental frequency. The only possible action associated with such elements is the generation of harmonics with respect to voltage (or current).*

Therefore, any fully rectified waveform that is "forcibly" returned to baseline (zero) at some point in the cycle will not display reactance since the stored energy in conservative elements is shorted to ground (baseline) at that point. Furthermore, if a bias voltage (DC offset) is applied to the electrical circuit as a whole (the tissue path) all that has been accomplished is that the baseline is shifted from 0 V to the new level relative to some reference. Hence, no current will flow across the cell membrane in response to such a bias voltage since both the external and internal cellular environments are referenced to the same, new baseline; the transmembrane potential remains at approximately -70 mV, internal to external, irrespective of the baseline level. To be effective, the bias must be created across the cell membrane from external to internal. The only way to accomplish this is by creating an electric field that varies from point to point in Volts/cm.

It will be seen that to observe a phase shift between voltage and current, the condition to be met is to couple the time-variant element with a source of current such that the time-variant resistance modulates a steady state of current flow.

The time-variant elements are the voltage-gated ion channels with their associated variable resistances. In the next quote, recall that I is current; hence an I—V plot is a current—voltage plot.

(In terms of the I-V plot this condition can be described by stating that the element is "biased" at an operating point and that a sinusoidal perturbation is impressed about this point.) In such a system, electrical energy is being degraded to Joule's heat at every instant throughout the cycle since the element is purely dissipative. The phase shift between voltage and current usually encountered in electrical systems arises from circuit elements that are conservative, namely, inductance and capacitance-the property of the circuit in this case has been defined as the impedance.

"Biasing" an electrical element means that a nonzero baseline voltage is present. Normally we think of the baseline voltage across a circuit as being 0 V such as when a switch is open and the voltage source (e.g., a battery) is disconnected from the circuit. If a resting voltage, a bias, is present and a signal is applied, that signal voltage is added or subtracted from the bias voltage baseline. This is shown in Fig. 3.7, discussed below.

The only ways to achieve this situation, of an applied electric field with imposed sinusoidal perturbation about the bias point, is to either purposely apply a bias voltage of known value or allow the baseline to "float" and not forcibly return the developed voltage to a baseline value, e.g., via grounding. Electrical engineers designing circuits apply best practices and work very hard to eliminate these floating "transient voltages" that can occur when waveforms are not returned to baseline. However, for the effects enabling Impedance Neurography, these transients are a necessity and this observation constitutes a new aspect of neuronal cell membrane biophysics. At appropriate frequencies, this approach allows the bias to develop as the conservative capacitive or inductive elements within the circuit store energy.

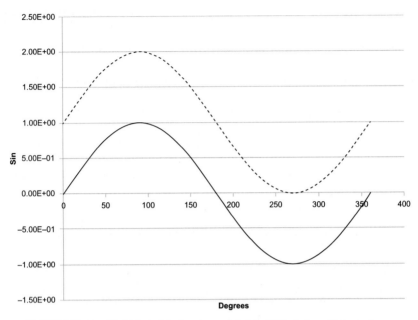

FIGURE 3.7 Rectified (*dashed line*) and unrectified (*solid line*) sinusoidal waveforms.

Consider that a fully rectified waveform has a nonzero average value (the dashed line in Fig. 3.7), as contrasted to an unrectified waveform that has an average value of zero (the solid line in Fig. 3.7). Unless grounded at 270 degrees, the result of a nonzero average voltage is that the conservative elements in the circuit store energy in the form of charge accumulation (capacitance) or magnetic field development (inductance). This will be clearly seen with the use of square waveforms as in Figs. 3.13 and 3.14 a bit later where these figures display a pulsed-DC waveform with resultant developed voltages. The voltages in those figures are depicted for circuits with RC time constants that are small, equal to, or large in comparison to the pulse duration, or $\frac{1}{2}$ the duty cycle (pulse on, pulse off) in the examples shown.

Before getting to Figs. 3.13 and 3.14, a discussion of what constitutes an RC time constant is in order. Recall that in the discussion of the cable equation that William Thomson and Oliver Heaviside derived, the time constant, represented by the Greek letter τ, was seen to modify the small changes in voltage occurring with small changes in time. In the case of neurons, the time constant is the RC time constant of the equivalent electrical circuit where RC refers to "Resistance" "Capacitance." Time constants are a way of describing how decay processes occur per equations in the form of Eq. (3.4):

$$C_t = C_0 e^{-kt} \tag{3.4}$$

where C_0 is the initial concentration, C_t is the concentration at time t, and k is a constant. When dealing with a time constant (τ) as in radioactive decay,

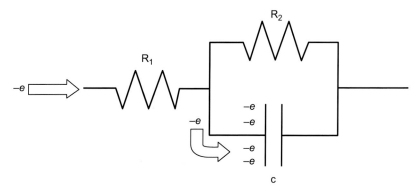

FIGURE 3.8 Displacement current flowing in a parallel RC circuit in the first few nanoseconds.

changes of gas concentrations in an anesthesia circuit, or the charging and discharging of an RC circuit in electronics, k is equal to $1/\tau$. If we want to see the charging of an RC circuit rather than the discharge, the applicable equation is of the form of Eq. (3.5):

$$C_t = C_0\left(1 - e^{-kt}\right) \tag{3.5}$$

How do these equations relate to nerve membrane responses in nerve stimulation?

Recall that the neuronal cell membrane may be represented, electrically, as a p-RC circuit as in Fig. 3.1. When a voltage is applied across an RC circuit, the initial few nanoseconds of current flow result in charges flowing onto the capacitive surface with no resistance: the so-called displacement current depicted in Fig. 3.8.

The consequence of charge flowing onto the capacitive surface without resistance is that no voltage develops across the circuit. This can be seen by using Ohm's Law (Eq. 1.1) and setting R to a zero value. Though the charge movement constitutes the displacement current flow, if R is zero, then E must also be zero. Very quickly, however, the accumulated charge on the capacitive surfaces begins to repulse additional charges coming on board and impedance to additional charge accumulation begins to occur. This mutual repulsion of like charges results in progressively more current being directed through the resistance of the p-RC circuit with the passage of time (Fig. 3.9).

Pushing the charges through the resistance portion of the circuit is an energy-requiring process and as more and more of the current is directed through the resistance, the developed voltage across the circuit rises. Finally, when the capacitive surface is completely charged, all the current is directed through the resistance of the p-RC circuit (Fig. 3.10) and the impedance becomes equal to the resistance. At this point, Ohm's Law could be applied to the circuit to calculate the voltage that would result from the applied current since no charge is flowing onto the capacitive surfaces. It is apparent that prior to the

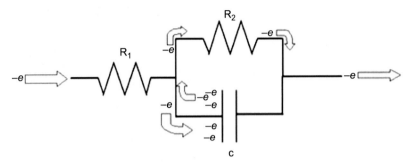

FIGURE 3.9 Current flow in a parallel RC circuit, a portion of which is being directed through the resistance of the circuit due to mutual repulsion of charges on the capacitive surface.

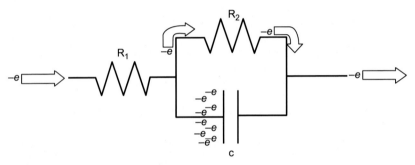

FIGURE 3.10 A fully charged parallel RC circuit with continued current flow directed entirely through the parallel resistance.

capacitive elements being fully charged, the impedance of the circuit is less than the resistance. Not realizing this made for some early confusion on my part when starting the investigations.

How does the charging equation relate to this process? Fig. 3.11 helps to explain by showing how the voltage develops.

We use the charging equation to describe the voltage difference (so-called voltage drop) across the circuit. From that equation, you can see that there is a time factor involved. This would lead one to expect that the voltage difference across the circuit varies in time as in Fig. 3.11.

A couple of important numbers can be derived from the parameters of the circuit components. It turns out that the time constant for the circuit equals the product of R and C. This is the amount of time it takes for the voltage to rise to within 37% of its final value; 0.37 being the value of 1/e. If this were a decay curve, the voltage would fall to 37% of its maximum in one time constant. And, since we are used to expressing decay in half-life (e.g., radioactive decay), the two parameters must be related somehow. They are; the half-life ($T_{1/2}$) equals the time constant times the natural logarithm of 2, as seen in Eq. (3.6).

$$T_{1/2} = \tau \ln(2) \tag{3.6}$$

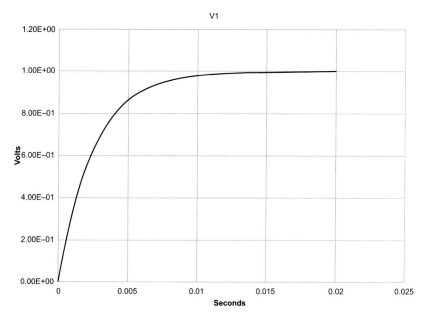

FIGURE 3.11   Developed voltage across a parallel RC circuit with a resistance 2500 Ohms and a capacitance of 1.0 μF. $\tau = 3.75$ ms. $T_{1/2} = 2.6$ ms.

In Fig. 3.11, I have chosen the resistance and capacitance to mimic those of the mammalian spinomotor neuron.[1] To calculate the RC time constant for a neuronal membrane it is necessary to use the capacitance per cm^2 and the membrane resistivity in Ohms·cm^2, which are the values used in Fig. 3.11.

If R and C are multiplied together, how is the result expressed in seconds? Consider that R (Ohms) is equal to V/Amps (remember Ohm's Law $E = IR \rightarrow R = E/I$) and that an Amp represents a Coulomb of charge passing a point every second, or Q/s. Then an Ohm is equivalent to a V·s/Q. A capacitor rated at 1 F will produce a potential difference of 1 V between its plates when 1 C of charge is stored. Therefore, a Farad is equivalent to Q/V. Multiplying R and C is (V·s/Q) × (Q/V) and equivalent to seconds.

I have been harping on the fact that a voltage gradient must preexist current flow, and yet in the example about which I have just been discussing, charge moves onto the capacitive surface with no change in voltage, i.e., the voltage differential is zero for the first few nanoseconds. Remember, in a DC circuit, current cannot flow across a capacitor. So, if a battery is connected across a p-RC circuit, the mutual repulsion of electrons sets up a local electric field and causes some of them to move onto the capacitive surface by migrating in that local electric field. There is no flow between the two poles of the battery during this early stage (Fig. 3.12) and the voltage differential across the circuit is zero.

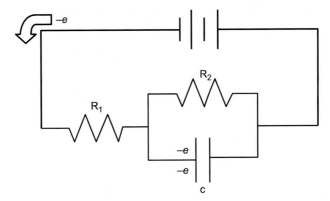

FIGURE 3.12 A battery connected across a parallel RC circuit for the first few nanoseconds. Note no electron flow returning to the battery as charge accumulates on the capacitive surface.

Once, current begins to flow through the resistance and return to the battery, the work of "pushing" charges through the resistance causes the voltage to fall across the circuit. This work is dissipated as heat; hence a resistor is termed a dissipative element. The capacitor, on the other hand, conserves charge and is termed a conservative element. By the way, this displacement current in a capacitor is similar to the gating current in a sodium channel where charges are simply moving from point A to point B, but not flowing between the electrodes, i.e., there is no current flow across the circuit. The movement of the charged moieties in the sodium channel can be thought of as similar to the aluminum foil movement in the demonstration of Coulomb's Law discussed in Chapter 1. In that case, there was no current flow between the foil and the charged plastic comb, but the foil moved because of the attractive forces acting on surface charges on the foil. In that example, the movement of the charges on the foil surface constituted current flow since the charges were moving in space. In the case of a voltage-gated channel, when those charged moieties move in response to the force of a voltage gradient, the molecular "stalk" on which they reside acts something like a lever to reposition parts of the larger molecule.

The Nature of a Charged-Direct Current Offset

In Fig. 3.13, the developed voltage for a circuit with a time constant equal to the current pulse duration rises to 63% of its predicted final value (1 in this case), corresponding with the definition of a time constant where $100\% - 63\% = 37\%$ or 1/e. By forcing the waveform to zero between pulses, which is consistent with good electrical engineering practice, the voltage always develops from zero with each successive pulse achieving 63% of predicted.

In Fig. 3.14, the voltage is not grounded between current pulses and declines, during the "off" portion of the duty cycle, on the exponential decay curve determined by its time constant using the maximum developed voltage at pulse termination as the starting point. Therefore, the voltage of the circuit

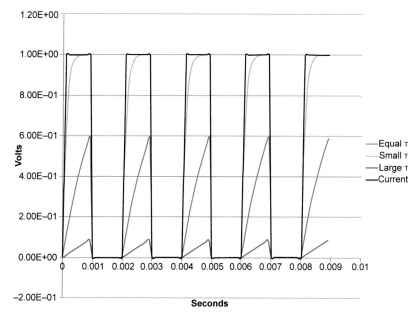

FIGURE 3.13   Rectified, uniphasic positive, grounded square waveform with resultant developed voltages for various circuit time constants. The *blue line* shows the developed voltage for a pulse duration that equals the circuit time constant, whereas the *red line* depicts the developed voltage for a pulse duration that is very short compared to the circuit time constant and the green line a pulse duration that is very large compared to the circuit time constant.

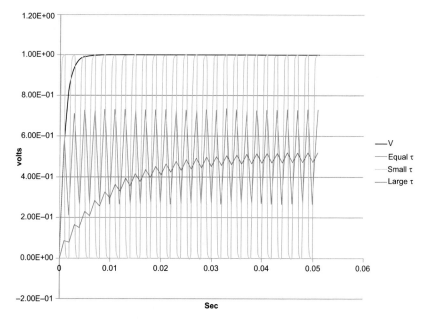

FIGURE 3.14   Rectified, uniphasic positive, ungrounded square waveform with resultant developed voltages for various circuit time constants as described in Fig. 3.13.

with a time constant equal to pulse duration at the initiation of the next current pulse will have declined only 63% of the previously developed voltage and will not have returned to zero. This effect is even more exaggerated for a circuit with a time constant very much larger than the pulse duration. Repeated pulses result in a rising voltage differential developing across the circuit that continues until the decay portion of the pulse equals the charging portion, and the waveform stabilizes. The residual voltage across the circuit at the end of each decay portion results in charge storage in the conservative elements of the circuit and a gradually changing baseline. For tissue near the anodal electrode, this change will be positive. Near the cathodal electrode, a voltage decline will occur, resulting in a net voltage differential in this example between the anode and cathode of 1 V (Fig. 3.15).

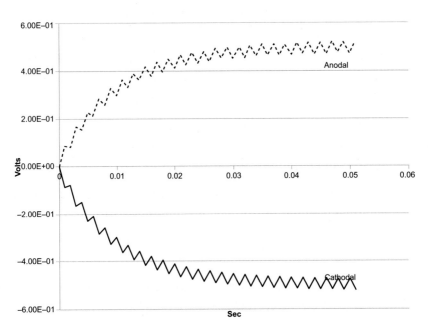

FIGURE 3.15 Voltage profiles developed at the anodal and cathodal electrodes for a circuit with a time constant much larger than the pulse duration.

The c-DC offset voltage that develops via the mechanism leading to the data in Fig. 3.14 is always $1/2$ the maximum at the cathode and anode predicted by the RC charging curve for a pulse duration that is very long compared to the circuit RC time constant. This is actually the average voltage for any periodic waveform applied to the circuit that is not forcibly returned to baseline between pulses, but is usually masked, or not considered, when high frequency waveforms, such as the green trace in Fig. 3.14, are employed and the peak-to-peak voltage differences are the desired parameter.

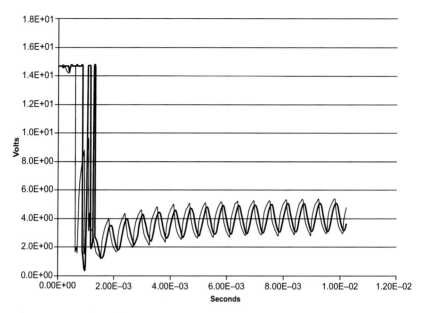

FIGURE 3.16 Fully rectified, square and sinusoidal waveforms at a frequency of 2 kHz, 6.7 mA amplitude, with floating baselines applied to a skin surface electrode overlying the course of a peripheral nerve. *(Disclosed in Cory, US Pat. Appl. 20,120,323,134.)*

This voltage differential is distributed over the circuit (tissue path) in Volts/ cm and does not represent merely a change in the baseline, but rather a gradient across the whole circuit including at the dimensions of the cell membrane. Such an external voltage gradient across the cell membrane promotes current flow and the criteria for resonance are met (imposed, nonzero periodic waveform on current flow across the conservative elements). Fig. 3.16 shows this effect from applying a square waveform and sinusoidal waveform to the skin surface overlying a peripheral nerve.

Now, having understood the mechanism underlying the c-DC offset, there are some consequences to discuss.

Consequences of Applying a Periodic Waveform on a Constant Current Flow Caused by a Charged-DC Offset

One consequence of a c-DC offset as described above is that transmembrane resting potential may be changed. This would be an expected outcome of a constant current flow being directed across the membrane, whereas normally there is no net current flow across the resting membrane, but only the regional microcurrents in both directions. Remember in the discussion of the ionic conductances of the neuronal cell membrane, the existence of microcurrents occurs as a consequence of the oppositely directed batteries that are associated

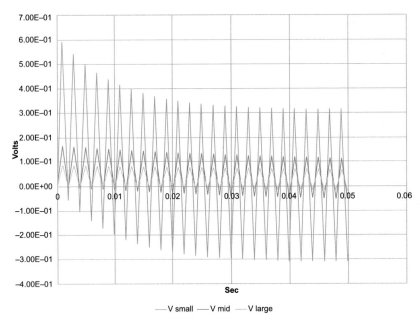

FIGURE 3.17 Voltage responses to biphasic-controlled current square waveform at 500 Hz per Raymond.

with the sodium, potassium, and leakage currents. Since those batteries are in opposition and there is the Na−K-ATPase assisting in maintaining the correct concentrations of ions across the membrane, no net macrocurrent exists in the resting state. With a c-DC offset, there is a resting macrocurrent because of the imposed bias voltage that is present, even at the level of the cell membrane. The consequence of the continuous macrocurrent is that an alteration of the resting cell membrane potential must exist. The thing is, do we have any kind of evidence supporting this?

Raymond discloses in US patent 5,775,331 that a more robust and rapid response to nerve stimulation occurs if "priming pulses" are first applied.[5] The system described in that patent produces a controlled current, biphasic, square waveform that does not appear to be grounded between pulses.

If a waveform conforming to the Raymond parameters (900 μs pulse width, 100 μs interpulse dwell time) is applied to tissue, the developed voltages depicted in Fig. 3.17 occur where V small refers to the developed voltage for a system time constant that is small (1 ms), V mid refers to a 5 ms time constant, and V large refers to a 10 ms time constant. What is apparent is that as the ratio of the time constant to the frequency increases, a more prolonged c-DC offset develops, which tends toward zero over time. However, if the ratio exceeds a critical value, the c-DC offset remains because the lowest developed voltage value does not drop below 0 V, even though the applied waveform may be $(+)$ A to $(-)$ A. This is seen in the V-mid and, even more so, in the V-large traces.

The basis for the observation of Raymond, that priming pulses allow for a more rapid and robust response to stimulation pulses, was not disclosed in the patent, but revealed another aspect of the sustained c-DC offset discussed above. The data reveal that the externally applied voltage required for depolarization and action potential formation was reduced. This is obvious as the sustained gradient acts to reduce the transmembrane potential difference. Since a 6–7 mV depolarization across the neuronal cell membrane is required in the resting state to cause action potential formation, a sustained c-DC offset results in a baseline depolarization, say 1–2 mV, less voltage differential is required of the stimulating pulse to effect action potential generation.

It is important to note that the c-DC offset described above is different from what is usually meant by that term. Typically, a DC offset refers to changing the baseline to which the circuit voltages are compared. Rather than maintaining a zero-voltage baseline, adding a DC offset is usually meant to indicate the baseline is now V volts rather than 0 V. Such a baseline change moves the whole electrical system V volts. In terms of the transmembrane voltage differential in a nerve cell, approximately -70 mV, adding a baseline DC offset does not change that transmembrane voltage differential. For instance, adding a 1 V DC offset to the transmembrane voltage differential changes it from 0.00 to -0.070 V outside to inside to 1.00 to 0.030 V outside to inside and maintains the voltage differential of a -70 mV. Adding the DC offset through the action of a floating baseline creates the offset through accumulation of charge on tissue capacitive surfaces. Such a DC offset may change the transmembrane voltage differential, say from approximately -70 to -68 mV.

It is instructive to analyze the electrical current path that occurs during Impedance Neurography and to consider the anisotropicity represented by nerve (Fig. 3.18).

Keratinized skin (Fig. 3.18) is represented as a p-RC element since no voltage-gated channels are present in that layer, the large nerve as a p-RLC element, and the small fibers as a second p-RLC' element. Since the skin and nerves are in series, the equivalent circuit of three parallel elements in series may be shown as in Fig. 3.19.

Due to the electrical anisotropy that living nerve displays in a developed c-DC offset field, as shown in Fig. 3.18 and discussed in Chapter 4, the bulk of the applied current flows along the nerves rather than distributing through the surrounding tissue. To test the correspondence of the model with the data from Fig. 3.4, several graphs of calculated data using physiologic parameters were constructed. First shown is a p-RC circuit (Fig. 3.20), then a p-RLC circuit (Fig. 3.21), and finally a series p-RC—p-RLC—p-RLC' circuit (Figs. 3.22 and 3.23).

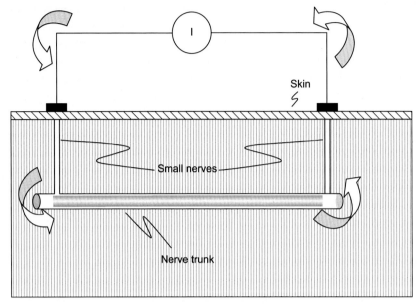

FIGURE 3.18 Representation of nerve electrical anisotropic pathway for current flow in tissue. *Arrows* show current flow preferentially along nerves.

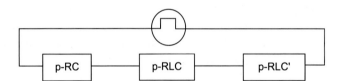

FIGURE 3.19 Schematic representation of the pathway shown in Fig. 3.12.

Calculation of the impedance of these circuits was conducted with the use of the complex impedance method. This is not a straightforward proposition, but uses the relationships shown in Eqs. (3.7)–(3.9):

$$Z_{eq} = R_{eq} + jX_{eq} \tag{3.7}$$

$$Z_{eq} = |Z|^{ej\theta} \tag{3.8}$$

$$Z_{eq} = \sqrt{\left(R_{eq}^2 + X_{eq}^2\right)^{ejarctan\left(\frac{X}{R}\right)}} \tag{3.9}$$

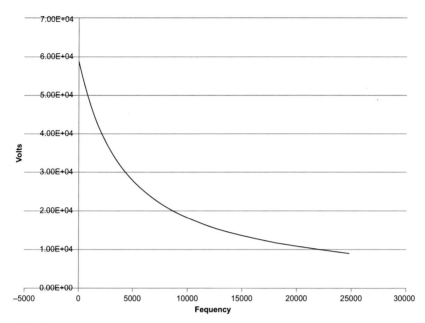

FIGURE 3.20   Z versus frequency for Impedance Neurography model as a parallel RC circuit.

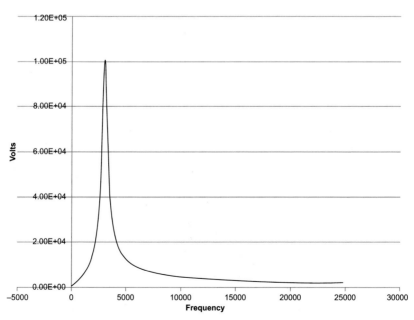

FIGURE 3.21   Z versus frequency for Impedance Neurography model as a parallel RLC circuit. Values for R, L, and C per Sabah and Leibovic.[10]

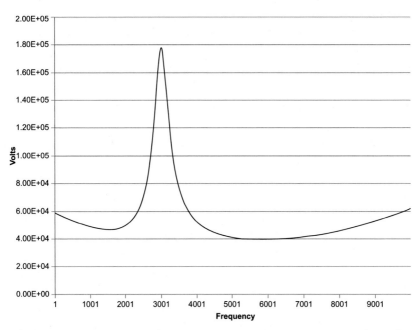

FIGURE 3.22 Z versus frequency for Impedance Neurography model as a series p-RC-p-RLC-p-RLC′ circuit.

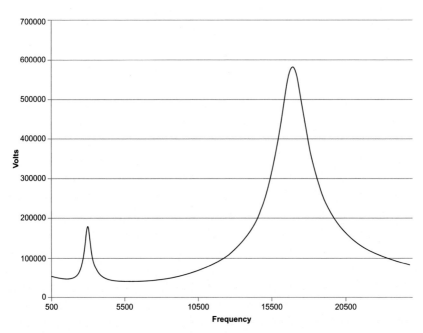

FIGURE 3.23 Z versus frequency graph for p-RC—p-RLC—p-RLC′ circuit carried out to greater than 20,000 Hz showing second resonance peak.

Contributing to the complexity of these calculations are the relationships defining R_{eq} and X_{eq} in parallel RLC circuits shown in Eqs. (3.10) and (3.11):

$$R_{eq} = \frac{((R_L R_C + L/C)(R_L + R_C) + (\omega L R_C - R_L/\omega C)(\omega L - 1/\omega C))}{\left((R_L + R_C)^2 + (\omega L - 1/\omega C)^2\right)}$$

(3.10)

$$X_{eq} = \frac{((\omega L R_C - R_L/\omega C)(R_L + R_C) - (R_L R_C + L/C)(\omega L - 1/\omega C))}{\left((R_L + R_C)^2 + (\omega L - 1/\omega C)^2\right)}$$

(3.11)

Fortunately, the phase angle, θ in Eq. (3.8), is fixed at between 70 and 80 degrees for biological membranes over a wide range of frequencies.[1] The fixed phase angle may seem a bit odd, and certainly did to the investigators who have yet to elucidate the underlying mechanism. Cole states this succinctly on page 42 of *Membranes, Ions and Impulses,*

It has since received many various experimental confirmations although, as everyone freely admits, no fundamental explanation for either the constant phase angle or the frequency dependence of the resistance or capacitive components for such an impedance has yet been found.[1]

Our work may explain this finding. The phase angle is that difference between the sine wave positions of current and voltage in an electrical circuit containing both dissipative (resistance) and conservative (reactance) components. To picture this, visualize the sine graphed against degrees. At 0 degrees, the sine is 0, while at 90 degrees the sine is 1. It returns to 0 at 180 degrees and is -1 at 270 degrees. Now, let us perform a *Gedankenexperiment*. Recall from our discussion of RC and RLC circuits that the development of a voltage across the circuit requires time related to the storage of energy in the form of charge (capacitance) or magnetic fields (inductance), both of which are not instantaneous processes. Therefore, the voltage curve will *lag* the current curve and the peak voltage will occur at a higher degree number than the current curve. If the curves are 90 degrees out of phase, the current curve will reach a normalized value of 1 at 90 degrees, but the voltage curve will not reach a normalized value of 1 until 180 degrees. The phase angle in this case is 90 degrees or the difference between the positions of the sine waves for current and voltage.

If the voltage-gated channels are behaving like variable capacitive and inductive elements, their responses to time-variant electrical waveforms could explain the fixed nature of the membrane phase angle. Usually, the membrane capacitance and resistivity are thought to be fixed values that relate to the lipid bilayer structure of the membrane itself, albeit with imbedded "holes" (channels) that render the overall structure leaky. What has not been considered in most, or perhaps all, treatments of these membrane electrical parameters are that variable components exist that can behave as capacitive or

inductive elements in addition to the lipid bilayer capacitance and resistivity. This is likely the explanation for Cole's observation of what he calls "anomalous impedance" and is well covered by Mauro as quoted earlier in this chapter.

The existence of a constant phase angle means that the exponential component of Eq. (3.9) is a constant and the shape of the curve is solely defined by the values of R_{eq} and X_{eq}. Also, the Greek letter ω in Eqs. (3.10) and (3.11) stands for the angular frequency and equals $2\pi f$. So, all that is required for calculation of the curves is a lot of algebra, the purpose for which computers were designed. Using parameters for membrane resistance, capacitance, and inductance from the literature resulted in the calculation of the curves in Figs. 3.20–3.23.[2,10]

Fig. 3.22 is the only representation that conforms to the experimental data shown in Fig. 3.4, even including the odd upsloping of the Z versus frequency graph at higher frequencies, as distinguished from the expected asymptotic approach to a minimum. This upslope actually presages another resonance peak associated with the p-RLC′ element (small fibers) shown in Fig. 3.18. Also, though Fig. 3.4 (experimental data) and Figs. 3.22 and 3.23 show relatively smooth curves, in fact for those data comprising Fig. 3.4, the "smoothness" was an artifact of the frequency sampling interval: 50 Hz in that case. At smaller sampling intervals of 1 or 2 Hz, the single peak devolved into a myriad of small, much more sharply defined peaks. The question remains unanswered as to whether the large number of component peaks were related to neuronal subpopulation responses. Notably, the resonance peaks in Fig. 3.23 occur at the same frequencies that have been empirically determined as effective for high frequency neuromodulation…there is a reason for that.

High frequency neuromodulation is a nerve stimulation technique applied either to the spinal cord or peripheral nerves for pain control. It has been observed to have mixed results with some reports indicating that it does work while others, particularly animal studies, have been less clear. What is apparent from the frequency studies with Impedance Neurography equipment is that electrical resonance, which would be expected to cause electrical oscillation of the neuronal cell membrane, occurs at the same frequencies found useful in these studies of high frequency neuromodulation. My suspicion is that though the setup in high frequency neuromodulation remains suboptimal, enough membrane oscillatory effect must occur to provide variable results on pain control. Below, I present a discussion of how such systems may be optimized. First, however, an aside prompted by the shape of the curves in Figs. 3.20–3.23.

Fig. 3.21 is interesting as it represents the situation of tissue without the p-RC component of the skin. From very early studies of Impedance Neurography, we noticed that intact skin was important for Impedance Neurography measurements and that discrimination of nerve tissue was lost when the skin was removed or disrupted by scratches. This was verified with sanding experiments where the subject (me) had a patch of skin sanded by an enthusiastic and thorough electrical engineer with 220 grit sandpaper to remove the keratinized layer. Then, surface impedance determinations were conducted with the Impedance Neurography equipment. Three things were obvious; (1) the formerly sensationless Impedance Neurography technique was now associated with marked pinprick sensations, (2) the discrimination of the underlying nerve structure was lost, and (3) you want to be careful just who you let sand your skin. Since both the p-RLC and p-RLC-p-RLC' circuits show the same very low impedance values at frequencies less than 2000 Hz, the reason for the observed loss of nerve discrimination is apparent. Also, since the high impedance (and the voltage drop across it) of the keratinized layer under the sampling electrode was lost, pinprick sensations (electric shock) could occur due to the larger voltage gradient being applied across the p-RLC-p-RLC' circuit.

An electrode system we considered, briefly, included micromachined spikes to puncture through the stratum corneum of the skin without poking deep enough to reach the dermis. That was some nifty technology, but inspection of Fig. 3.21 shows that it would not have benefited Impedance Neurography determinations because loss of the p-RC component of the keratinized stratum corneum would eliminate the initial, low frequency impedance separations that were seen between high nerve density tissue and low nerve density tissue.

Fig. 3.21 also provides a clue for managing Impedance Neurography determinations when no keratinized layer is present to provide the p-RC element, e.g., in a surgical field such as locating the facial nerve during parotid gland surgery. Since the parallel elements of the circuit in Fig. 3.19 are in series, the p-RC portion of the circuit could be supplied by an external circuit with parameters closely matched to the keratinized layer resistance and capacitance. In fact, such an external circuit in series with the tissue circuit may be advantageous for optimizing the impedance differences between high nerve density tissue and low nerve density tissue. In other words, "tuning" of the effect may be possible using such an external circuit wherein the added series p-RC element has adjustable resistance and capacitance components.

The essential point from the foregoing discussion is that to observe resonance an absolute necessity exists for an external current flowing across the neuronal cell membrane that leads to the changes in time-variant resistance of the ionic conductances. *Establishing such a transmembrane current is only possible if a developed c-DC offset is present in the applied, external field such that voltage gradients are created across the membrane rather than simply shifting the baseline up or down.*

Recall that voltage gradients are created exclusive of charge movement. This is the realm of electrostatics, shown in Chapter 1 using the Coulomb's Law example describing the electric field in Volts per unit distance at any point between two fixed charges in space, shown again in Eq. (3.12).

$$F = K(Q_1 Q_2 / d^2) \qquad (3.12)$$

That voltage gradients are established across neuronal cell membranes by remotely applied voltages is apparent from the ability to generate action potentials in neurons suspended in oil baths. Since oil is an excellent insulator, no current flows when a voltage is applied between electrodes suspended in the oil, but with sufficient step voltage magnitude, action potential generation occurs. Mathematical formalization of the effects of externally applied fields can be found in Cooper.[6] For DC currents, Cooper demonstrates that voltage gradients of 100 mV/cm with pulse durations of greater than 0.5 times the membrane time constant in the vicinity of the neuronal cell membrane are required to create transmembrane gradients of 6—7 mV necessary for action potential generation.

Impedance Neurography employed field strengths much smaller than that required for action potential generation, but nonetheless sufficient to create transmembrane voltage gradients and consequent current flow. Since the field was time variant (sinusoidal), all the necessary criteria were met for creation of the ionic conductance-related reactances and resonance at the critical frequencies. If, however, a sinusoidal field without a developed c-DC offset was applied to tissue containing nerve, no such transmembrane current flow occurred and the ionic conductance-related reactances were not observed. This is the case with most impedance studies of tissue where good engineering practice returns the waveform to zero or a nonzero baseline through grounding at the minimum values of the waveform. The grounding eliminates the reactances that may have developed in the tissue circuit, e.g., with a commercial gain-phase analyzer. Not appreciating this fact led at least one potential collaboration to go off the rails.

A group working in the Impedance Tomography area was interested in our findings and, following nondisclosure agreements between us, discussions had proceeded to the point that I revealed the necessity of a DC offset for our observations. I did not discuss just how we accomplished the point-to-point DC offset and the other group decided to check our findings using their commercial gain-phase analyzer, which, of course, demonstrated no resonance effect. Discussions abruptly stopped due to what was a great example of "ready-fire-aim."

These effects are shown diagrammatically in the following illustrations.

In Fig. 3.24, a tissue cross section through a neuron is shown with the cell interior and exterior marked. An externally applied baseline voltage level is depicted by the dashed line. Note that no voltage gradient exists across the cell membrane and hence no current flow is possible.

In Fig. 3.25, a square waveform, shaped by the RC circuit parameters, has been imposed on the baseline voltage level, each pulse varying from baseline to V volts. The $\frac{1}{2}$ cycle time is either > five system time constants, allowing the decay portion of the pulse to return to baseline before the next pulse occurs, or the developed voltage is forcibly returned to baseline at the end of each cycle. Consequently, though the waveform oscillates up and down from

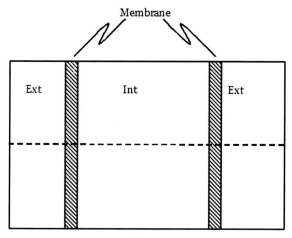

FIGURE 3.24 Diagram representing a block of tissue containing an individual neuron. The external (Ext) and internal (Int) regions of the neuron are marked and an externally applied, baseline voltage level is indicted by the *dashed line*.

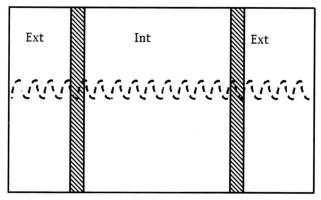

FIGURE 3.25 The tissue block of Fig. 3.24 with an imposed square waveform. *Int*, internal; *Ext*, external.

the baseline, there is no sustained voltage gradient developed across the cell membrane and no current flow. What is seen is essentially harmonic oscillation as Mauro described on page 51 above.

Fig. 3.26 depicts a tissue block in which the baseline voltage level has been increased above that of Fig. 3.25. However, the imposed waveform still does not create any voltage gradient across the membranes and therefore, there is no associated transmembrane current flow. In both Figs. 3.25 and 3.26, if the individual pulse amplitudes are large enough, depolarization of the membrane will occur with consequent action potential generation. Transmembrane

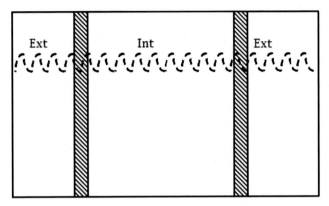

FIGURE 3.26 The tissue block of Fig. 3.25 with an increased baseline square waveform. *Int*, internal; *Ext*, external.

current flow is not necessary for this but is contingent on voltage-gated channels opening in response to the pulse voltage amplitude and ions flowing along the electrochemical gradient as described by the Nernst equation.

Fig. 3.27 shows the situation that arises when the electrodes are in close proximity to the neuron, whereas in Figs. 3.24–3.26, the electrodes have been remote. The pulse direction will be opposite at the opposite sides of the preparation, i.e., when the voltage deflection is 0 to (+) V/2 at the anode, it will be 0 to (−) V/2 at the cathode for a total voltage differential of V volts. Note, however, that no sustained voltage gradient exists across the membranes and no current flow occurs. The same effect related to pulse amplitude as discussed for Fig. 3.26 does pertain and if the magnitude of any given pulse exceeds the critical transmembrane voltage gradient of 6–7 mV, action

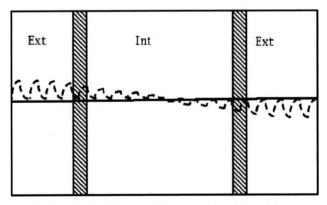

FIGURE 3.27 The tissue block of Fig. 3.24 with the anode and cathode in close proximity to the membrane surfaces. *Int*, internal; *Ext*, external.

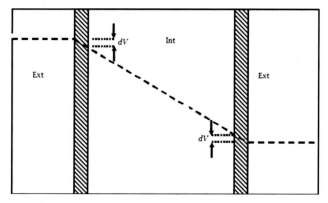

FIGURE 3.28 The tissue block of Fig. 3.24 with a transcellular direct current offset voltage gradient. *Int*, internal; *Ext*, external.

potential generation will occur and the nerve will fire. To create transmembrane current flow, a situation such as shown in Fig. 3.28 must exist.

In Fig. 3.28, the electric field is being created between electrodes in close proximity to the membrane, but it's easy to visualize the same effect as electrodes are removed from direct apposition to the membrane surfaces so long as the neuron lies between the electrodes.

The voltage gradient, dV, across the membrane causes current flow across the membrane. However, in contrast to the situation of electrodes in contact with the neuronal cell membrane, establishing such a gradient with electrodes applied remote from the position of the neuron is impossible using constant baseline waveforms due to the voltage distribution as seen in Figs. 3.25–3.27. This problem can be overcome by allowing the baseline to "float" as depicted in Fig. 3.16. By not forcing the voltage to a baseline value between cycles, the baseline follows a p-RC or p-RLC charging curve and establishes a c-DC voltage gradient across the preparation as conservative tissue electrical elements store energy, even to the scale of individual neurons. Consequently, as long as the field is applied, an externally imposed, transmembrane c-DC offset is present and any periodic field is imposed on that current flow (Fig. 3.29). The stage is set for the development of voltage-gated ion channel reactances that mimic both capacitive and inductive reactances and resonance becomes a possibility.

IMPROVEMENTS TO HIGH FREQUENCY STIMULATION

It is the lack of a c-DC offset that renders high frequency nerve stimulation nonoptimal as presently applied. Modeling based on standard equations describing depolarization (the Hodgkin–Huxley model of the axon) indicates that propagated action potentials will be blocked by membrane electrical

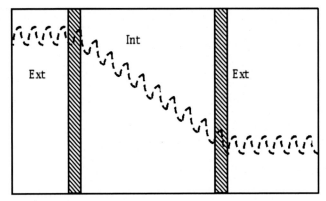

FIGURE 3.29 A periodic waveform imposed over a direct current offset voltage gradient. *Int*, internal; *Ext*, external.

oscillation at frequencies greater than 4 kHz, or as demonstrated above, at resonant frequencies. Both experimental and modeling work suggest that there exist optimal frequencies for this effect and that at frequencies above 20 kHz, the effect is minimal. Single axon studies confirm this.[7] The difference between single axon studies and externally applied fields to blocks of tissue, e.g., peripheral nerve stimulation or spinal cord stimulation, is that the imposed frequency cannot be selectively limited to the neuronal cell membrane in the block tissue example, whereas it can if one is dealing with a single neuron prep. The tacit assumption in using Hodgkin—Huxley equation modeling is that a single neuron is being assessed, and not a block of tissue. In a block of tissue, the whole tissue path for the electrical field will be oscillated when subjected to a periodic waveform, not just the membrane. This results in voltage harmonics for the tissue system, but the local gradient across any tissue structure, including the neuronal cell membrane, does not change. To be effective, high frequency stimulation must be able to exclusively drive oscillation of the neuronal cell membrane. This occurs naturally at the membrane resonant frequencies, which correspond to the empirically derived optimums from single axon preps and mathematical models. Importantly, these resonant frequencies will change with changes in the neuronal cell membrane physical/electrical characteristics, e.g., neuropathic changes such as diabetic neuropathy and neuronal subpopulation size differences.

DISCRIMINATING NEURONAL SUBPOPULATIONS

Two lines of evidence suggest that it is possible to discriminate neuronal subpopulations. One has been mentioned regarding sampling frequencies at shorter intervals and observing a multitude of peaks develop in the region of the main peak of Fig. 3.4. The other goes back to the confusing inability to image the median nerve with parameters that allowed imaging of the saphenous nerve.

The median nerve is a mixed function nerve containing both sensory and motor elements, whereas the saphenous nerve is devoid of motor elements; it is sensory specific. Using a sampling frequency of 2000 Hz, it proved impossible to image the median nerve as a low impedance structure. The explanation turned out to be that the resonance peak was already beginning to occur at that frequency, whereas imaging the saphenous nerve at 2000 Hz was perfectly doable as its resonance peak developed at higher frequencies. The two nerves, with different constituent neuronal subpopulations, displayed different resonance peak frequencies. Those observations raise an exciting possibility, that of imaging neuronal subpopulations or assessing state of activation of neuronal subpopulations with Impedance Neurography technology.

Production of electrical oscillation of the membrane results from the p-RLC circuit characteristics that occur with a transmembrane current flow, only possible with a developed c-DC offset waveform. Also, the magnitude of the c-DC offset at any point on the membrane depends on the geometry of the situation as shown in Fig. 3.30.

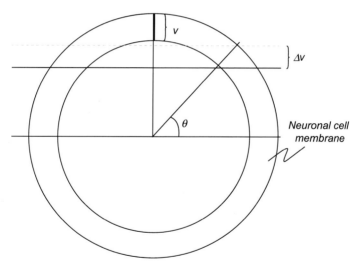

FIGURE 3.30 Cross section of a neuron in a uniform electrical field causing voltage gradient V at 90 degrees.

Fig. 3.30 represents a cross section of a neuron in a uniform electrical field, i.e., the voltage gradient is constant across the whole of the neuron (not attainable in fact because of distance factors, but useful for an illustration). At a position in the neuronal cell membrane where the voltage gradient is parallel to the slope of the membrane surface (90 and 270 degrees in this example), the transmembrane voltage gradient is $|V|$ ($+V$ or $-V$ depending on the side of the neuron being assessed). At all other positions in the cell membrane the transmembrane voltage gradient is proportional to $|V|\sin\theta$. Thus, the gradient is zero at 180 and 360 degrees. This means that the gradient vector will be oppositely directed across 50% of the membrane and that the same ionic conductance-related reactances will be capacitance-like in 50% of the membrane and inductance-like in the other 50% because of the orientation of the gradient vector.

The consequence of the profile of reactances in the membrane, subjected to a transmembrane potential gradient, is that the profile will be the same no matter the polarity of the field, leading to a constant, integrated reactance profile. The magnitude of the reactances will vary from point to point along the neuron depending on its orientation to the electrodes supplying the gradient and because in the clinical situation, the voltage gradient is not uniform (varying with distance).

The Relationship of the Time Constant to the Apparent Impedance

An additional effect of the inductance-like reactance is that the total tissue time constant changes. Examination of the initial portion of the p-RC curve (Fig. 3.20) and the p-RC-p-RLC-p-RLC′ curve (Figs. 3.22 and 3.23) demonstrates this effect; the p-RC curve shows a steeper impedance decline than do the p-RLC curves with increasing frequency. This time constant change is reflected in the way that impedance is calculated with the Impedance Neurography technology.

Impedance, like resistance, is a way of describing the current—voltage relationship across a circuit, or elements in a circuit. The resistance is defined as the voltage/current ratio, while impedance, using periodic waveforms, includes not only the voltage/current ratio but also factors in the phase angle. Practically speaking, impedance determination requires measuring the peak-to-peak voltage difference from the applied waveform whether using a controlled current or controlled voltage approach. These voltage measurements can be markedly affected by the circuit time constant as seen graphically in Figs. 3.31 and 3.32.

The data used to construct these graphs were derived from matrices of resistance and capacitance values that allowed choosing values for R and C that gave similar calculated impedances at 2000 Hz with differing time constants (Fig. 3.32), or similar time constants with differing impedances (Fig. 3.31).

Note that in Fig. 3.31, though the impedances vary by a factor of 2.5, the peak-to-peak voltage differentials are similar as are the circuit time constants. In Fig. 3.32, the impedances are similar, but the time constants vary by a factor of 3.2 with much bigger differences developing in the peak-to-peak voltage differences. In Figs. 3.31 and 3.32, the blue V line shows the charging curve for a p-RC circuit having the same time constant as the cycle time of the 2 kHz waveform: 0.0004 s. If a single, DC pulse of prolonged duration were to be applied to a p-RC circuit with that time constant, the circuit would develop a

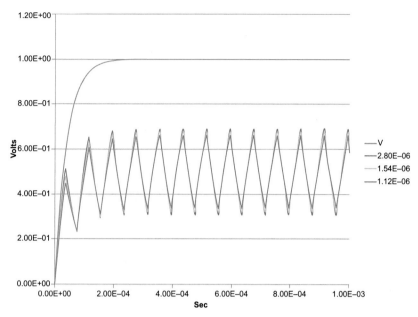

FIGURE 3.31 Impedance versus frequency results for three parallel RC circuits with similar time constants (5.00×10^{-5} s to 6.00×10^{-5} s) and impedances varying by 2.5-fold (2 kHz) when assessed by a controlled current, uniphasic, nongrounded, pulsed-direct current periodic waveform.

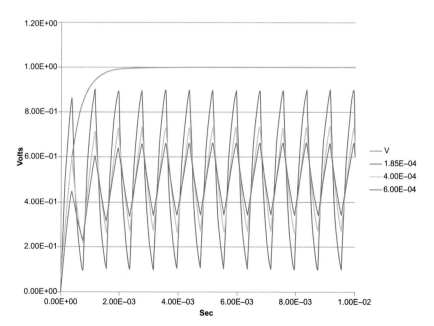

FIGURE 3.32 Impedance versus frequency results for three parallel RC circuits with similar impedances at 2 kHz ($3.18 \times 10^2\ \Omega$ to $3.43 \times 10^2\ \Omega$) and time constants varying by 3.2-fold when assessed by a controlled current, uniphasic, nongrounded, pulsed-direct current periodic waveform.

normalized voltage corresponding to the blue trace. Note that in both cases, the time-averaged voltage for the stabilized waveforms is 0.5 V, but in Fig. 3.31, both the peak-to-peak voltage difference and mean voltage are very similar plus, since the curves tend to track very closely, the phase angle with the current pulse will be very similar. Taking measurements with these parameters will provide very similar impedance determinations, despite the calculated values based on the resistive and capacitive components showing a 2.5-fold difference.

Now, consider Fig. 3.32. Once again, the mean voltage is 0.5 V, but in the face of widely varying time constants, the peak-to-peak voltage values are quite different. Also, since the traces are not tracking each other, it is clear that the phase angles will be different.

The conclusion from Figs. 3.31 and 3.32 is that the total tissue time constant is playing a very large role in the impedance value derived from the waveform peak-to-peak voltage measurements. Examination of Figs. 3.31 and 3.32 reveals that when using sampling waveforms with one-half cycle times that approximate the system time constant, much greater effect will be seen from variation in the time constant than variation in the impedance. Since nerve tissue time constants show electric field-dependence as discussed above, Impedance Neurography is a technique the uniquely takes advantage of this biophysical phenomenon.

A consideration that may not have occurred to those unfamiliar with the issues of determining these peak-to-peak values involves identifying the peaks and troughs to perform the calculation. A number of individual waveform cycles must be sampled at an interval that will be likely to select both the maximum and minimum values at some point. If the sampling rate is too low, a problem called "temporal aliasing" can occur. The Impedance Neurography equipment used to generate Figs. 3.31 and 3.32 sampled each waveform 200 times and the maximum and minimum values were obtained from the resulting dataset. When using equipment that performs such feats, most of us do not give any thought to what is occurring in the "black box" on which we depend for accurate answers. Then, there is the whole issue of how to display the information. Should one use a linear graph, a 3-D surface, the raw data, or manipulated data? When trying to explain these considerations to a medical scientist naïve to the nuts and bolts of computer use, his response was, "Just let the computer do it." Clearly, though we like to think of ourselves as distant from magical thinking, it sometimes pops out when we least expect it.

NEURONAL MEMBRANE ELECTRICAL CIRCUIT CHARACTERISTICS AND ANISOTROPICITY

Finally, the recognition of the neuronal cell membrane functioning as an p-RLC circuit offers a partial explanation to a question that has long been puzzling; how do nerves function as such effective anisotropic electrical

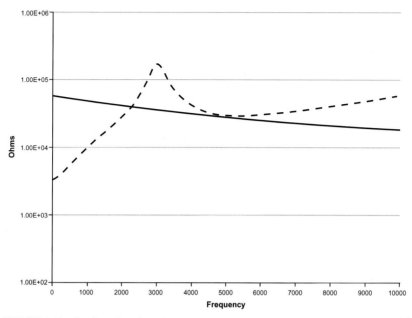

FIGURE 3.33   Semilog plot of parallel RC (*solid line*) circuit impedances and p-RLC-p-RLC′ (*dashed line*) circuit impedances. R $= 1000\,\Omega$, R′ $= 750\,\Omega$, C $= 4.0 \times 10^{-9}$ F, C′ $= 6.0 \times 10^{-10}$ F, L $= 0.70$H, and L′ $= 0.15$H.

features in tissue? Cooper et al. demonstrates that long, uninterrupted tubes can function as conduits for current.[8,9] But if the equivalent circuit was merely a p-RC circuit, the effect would have to depend on the magnitude of the capacitance alone since the intracellular and extracellular resistivities of neurons are similar to those of other cell types. The impedances of a p-RC circuit and p-RLC-p-RLC′ circuit show the answer as seen, graphically, in Fig. 3.33.

Fig. 3.33, a semilog plot of impedances for a p-RC and p-RLC-p-RLC′ circuits using physiologic values for R, L, and C shows the effect of the ionic conductance associated reactances for both large and small nerves.[10] The impedances of the p-RLC-p-RLC′ circuit are significantly less (0.5−2 orders of magnitude) than those of the p-RC circuit for frequencies under 5000 Hz with the exception of the region of the resonance peak. At frequencies less than the resonant frequency where optimal Impedance Neurography conditions have been determined experimentally, the impedance differences are most marked. It is these voltage differences and the consequent calculated impedance differences combined with the anatomy of the neuron that constitutes the electrical anisotropy of nerves. Hence, with the use of c-DC offset, nongrounded periodic waveforms, nerves become effective paths of least resistance for current flow, allowing visualization of their course and position. The requirement for the ionic conductance-related reactances also explains the necessity of living tissue for the effect since even freshly cadaveric tissue will display the membrane lipid bilayer capacitance and intracellular resistivity of living cells but lacks the channel electrical characteristics as cellular mechanisms required for their maintenance, e.g., ATP formation, are absent. Also, it is the density of voltage-gated channels in neurons that make them respond as low impedance structures in appropriately structured electrical fields, and why other cell types, e.g., cancer cells, with known high capacitance do not show up on Impedance Neurography (they are not paths of least resistance).

THE STRENGTH–DURATION RELATIONSHIP

Now we are ready to answer a question that puzzled me for years—the source of strength–duration curves. In textbooks, these curves seemingly appear out of nowhere and are presented as being somewhat mysterious. Their history goes back to around 1901 and the early studies using current pulses to depolarize neurons that were discussed earlier. Consideration of some of the parameters required for transmembrane voltage gradient generation reveals the explanation of strength–duration curves to be straightforward. Fig. 3.34 depicts a typical strength–duration curve.

Strength–duration curves can be determined using either current or voltage pulses. What they show is that as the duration of a stimulus pulse increases, the amplitude required to depolarize a nerve decreases to a minimum, nonzero value. This value even has a name; it's called the rheobase. That odd term comes from the Greek and means the foundation current referring to an infinite duration current of just sufficient magnitude that would cause a neuron to depolarize. That is why the

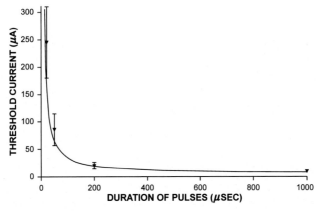

FIGURE 3.34 Strength–duration curve showing the duration of a stimulus pulse of various amplitudes required for nerve depolarization.

strength–duration curve is asymptotic and never quite reaches the minimum value. The question is, what accounts for this relationship between stimulus duration and stimulus amplitude; Fig. 3.5 provides a clue.

We know that the important factor for nerve depolarization is the voltage gradient that develops across the membrane. Effectively, then, to construct a strength–duration curve, the neuron is being used as a transmembrane voltage indicator showing when the voltage gradient across the neuronal cell membrane reaches the critical value and an action potential is set off. Since the membrane acts as a p-RC circuit during a single DC pulse (not a p-RLC circuit since no sustained c-DC offset develops with only one pulse), the transmembrane voltage gradient will rise exponentially according to the characteristic charging curve for that particular neuronal cell membrane. There is no way to know what final voltage would be achieved across the membrane for any individual current because the neuron fires long before that voltage level is achieved, i.e., it fires when the transmembrane potential gradient reaches the 6–7 mV level. At the point of action potential development, the voltage characteristics of the membrane become very nonlinear and describing them is a complex proposition. Fortunately, we do not have to do that because all a strength–duration curve relates to is how the trans-membrane voltage develops during that initial charging process. So, in effect, what we have done is to use a specific transmembrane voltage detector (the neuron) to determine just how long it takes at a given stimulus duration to climb up the charging curve to the depolarization voltage value.

The similarity of the curves in Figs. 3.34 and 3.35 is obvious. Fig. 3.35 is calculated using information one can surmise from the above rationale, not the

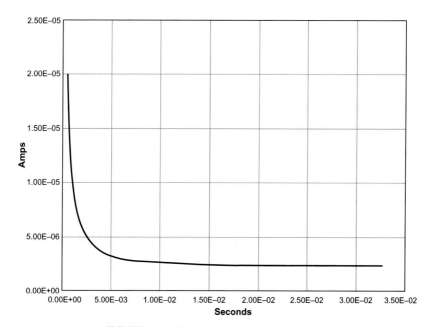

FIGURE 3.35   Calculated strength–duration curve.

usual Lapicque or Weiss equations, which attempt to use charge or current to explain the curve (you can look them up if you are interested). We already know from previous discussion that neither current nor charge is important factor in depolarization, but that the voltage gradient is the important factor. So how does that knowledge inform calculation of the strength—duration curve? In his 1995 paper, Cooper demonstrates that because of the p-RC nature of the neuronal cell membrane, the transmembrane voltage gradient will develop similarly to the information in Fig. 3.11, i.e., a decay (charging) curve function.

Here is how the curve in Fig. 3.35 was calculated. Noting the similarity to Eq. (3.4), the decay curve may be modified to a charging curve and show developed voltage as follows in Eq. (3.13):

$$V_t = V_0\left(1 - e^{-kt}\right) \tag{3.13}$$

Rearrangement gives results in Eq. (3.14):

$$\frac{\ln(1 - V_t/V_0)}{-k} = t \tag{3.14}$$

where time t is the pulse duration. Now, using R = 2500 Ω and C = 1.5 μF, we can determine k as $1/RC = 266.67\ \text{s}^{-1}$. V_t is set to 6 mV, my assumed transmembrane voltage gradient required for depolarization in this idealized neuron, and V_0 can be set to any value desired (so long as it's greater than 6 mV).

To determine the current, we assume the capacitor of the p-RC circuit is fully charged and all the current flows through the resistor. By setting the final voltage (V_0) to the desired value, we can determine the current from Ohm's Law, E = IR ($I = V_0/R$), since we know both E and R. Now, with the rearranged charging curve equation we are set to determine the pulse duration to achieve a V_t of 6 mV. The data follow in Table 3.1.

Do we have the unique current values for this idealized neuron, or can a number of different currents give the same result? What if the resistance and capacitance changed by a factor of 10: resistance 10× greater, capacitance 10× less? The first thing is that the time constant is exactly the same since it's the product of R and C. That implies that the charging curve will also be *exactly* the same. Something must change, however, and that something is the current. The strength—duration curve for the new parameters is shown in Fig. 3.36.

Notice that Figs. 3.35 and 3.36 look the same and that the pulse durations (seconds) are the same. In fact, the current is the same numeric value, just differing by a factor of 10. This is because the resistance portion of the RC circuit is now 10× greater so that for the second idealized neuron, 10× less current will be required to achieve the same transmembrane voltage gradient: the same transmembrane voltage gradient with one-tenth of the current.

This is what Cooper shows us—that the time to achieve an adequate transmembrane voltage gradient is dependent on the p-RC response of the neuronal cell membrane when exposed to an externally applied electric field in Volts/cm.

TABLE 3.1 Calculated data from the rearranged charging curve equation for Fig. 3.10

V_0	I	Time to 6 mV
0.006001	2.4004E-06	3.26E-02
0.00601	2.4040E-06	2.40E-02
0.0061	2.44E-06	1.54E-02
0.007	2.80E-06	7.30E-03
0.008	3.20E-06	5.20E-03
0.009	3.60E-06	4.12E-03
0.01	4.00E-06	3.44E-03
0.011	4.40E-06	2.96E-03
0.012	4.80E-06	2.60E-03
0.013	5.20E-06	2.32E-03
0.014	5.60E-06	2.10E-03
0.015	6.00E-06	1.92E-03
0.016	6.40E-06	1.76E-03
0.017	6.80E-06	1.63E-03
0.018	7.20E-06	1.50E-03
0.019	7.60E-06	1.42E-03
0.02	8.00E-06	1.34E-03
0.021	8.40E-06	1.26E-03
0.022	8.80E-06	1.19E-03
0.023	9.20E-06	1.13E-03
0.024	9.60E-06	1.08E-03
0.025	1.00E-05	1.03E-03
0.026	1.04E-05	9.84E-04
0.027	1.08E-05	9.42E-04
0.028	1.12E-05	9.04E-04
0.029	1.16E-05	8.69E-04
0.03	120E-05	8.37E-04
0.035	1.40E-05	7.05E-04
0.04	1.60E-05	6.09E-04
0.045	1.80E-05	5.37E-04
0.05	2.00E-05	4.79E-04

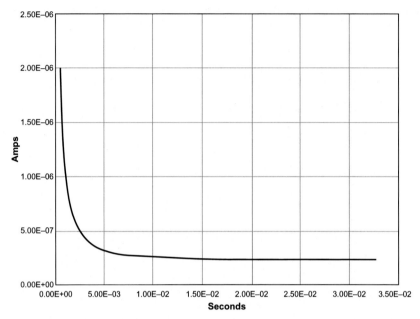

FIGURE 3.36 Calculated strength–duration curve using 10× greater resistance and 10× less capacitance.

It does not depend on charge accumulation or current as the Lapicque and Weiss equations indicate but is simply a function of the p-RC charging curve. And…the classic strength–duration curve can be easily derived from those voltage relationships. These curves are effectively showing which current · pulse duration is associated with a constant membrane depolarization voltage.

INFORMED NERVE STIMULATOR DESIGN

An additional thing to consider from the foregoing is adequate pulse duration for nerve stimulation.

Pulse durations of 200 μs are standard in nerve stimulators. Some have the capability of providing shorter pulses (100 μs) and may go up to 1500 μs. One of the things that Cooper's paper shows us is that to be effective, a minimal amplitude pulse duration must be 0.5 times the membrane time constant. Mammalian spinomotor neurons have time constants that range from around 3 to 15 ms.[2] It is obvious that industry standard nerve stimulators delivering pulses of 200 μs only achieve durations that are, at best, 0.067 times the shortest applicable membrane time constant (3 ms). As a consequence, these commercial nerve stimulators must drive output amplitudes much in excess of minimal values to impel the voltage adequately up the charging curve in the time of the pulse duration as shown in Fig. 3.5. The argument for these short durations comes from some historical work on dorsal column fibers in the spinal cord and not peripheral spinomotor neurons.[11] Plus, the sentiment that longer duration pulses will activate sensory neurons and be painful is often used as a reason for shorter duration pulses. Not only does this reflect a nonunderstanding of the development of similar transmembrane voltage gradients in the various neuronal subtypes based on the dimensions and constituents of the membranes but also forgets basic neurophysiology that recognizes sensory information being encoded both spatially and temporally. Spatial encoding refers to the fact that small diameter fibers transmit pain information, and temporal encoding refers to the requirement for stimulation frequencies in excess of 15 Hz to be perceived as painful. Low frequency, long pulse duration stimuli, e.g., 5 Hz, are perceived as tapping sensations even if the fibers being activated are C-fibers or $A\delta$-fibers (small diameter, nociceptive fibers). These short duration pulses only become painful if the amplitude is turned up to the point that sufficient motor fibers are recruited to fire and lead to painful muscle contractions. At minimally effective amplitudes, used for accurate nerve location, such painful contractions do not occur.

Then there is the unfortunate issue of lore in nerve stimulator design. Many sources, particularly in the anesthesia literature, stress the importance of nerve stimulators being controlled/constant current devices. The thinking is as follows: since delivery of adequate current is necessary for effective nerve stimulation, stimulators should have controlled current outputs to compensate for changes in system impedance that may occur. This point of view is uninformed for two reasons, both of which have been discussed earlier.

First and foremost is the fact that current is irrelevant to nerve stimulation. Hopefully, I have reiterated that historical point sufficiently.

Second, since the developed transmembrane voltage gradient is of primary importance for nerve cell membrane depolarization, designing nerve stimulators to deliver a constant current output means that as the system impedance falls the developed voltage necessary to drive the constant current will also fall (it takes less voltage force to drive the same current through declining impedance). The fall in output voltage coincides with a concomitant fall in the transmembrane voltage gradient and loss of the ability to depolarize the nerve cell membrane. With increasing length of stimulating needle or catheter insertion, impedance will decline as I have discussed and documented earlier. Furthermore, there is no mechanism to increase system impedance with needle insertion as it is at its maximum when the needle first touches the skin surface. From there on, with increasing depth of insertion, the impedance can only fall due to the variable capacitance to which I have alluded. Consequently, forming an adequate voltage gradient sufficient to depolarize a neuron will be impaired when using controlled current as the needle depth is increased. The only way around the problem of adequate voltage gradient maintenance is to use controlled voltage output nerve stimulators. Though local voltage gradients will vary depending on the nonhomogeneity of the surrounding tissue, the voltage gradients will not be subject to system impedance—related changes since the distance between the cathode (needle) and anode (return electrode) is essentially unchanged. To visualize this, think of the field distribution lines in Fig. 1.5. That Volts/cm distribution pattern in the block of material is independent of the resistance (or impedance) of the intervening material, only depending on the separation of the electrodes in space.

Some nerve stimulation devices do indeed use controlled voltage outputs. Notably, these are usually found on radiofrequency ablation units, which are very effective at allowing ablation electrodes to be placed in proximity to the target nerve. Hopefully, as design electrical engineers read the foregoing material, more controlled voltage nerve stimulators will begin to appear in the market enabling more precise nerve location, especially for those nerves located beyond the depth limitations of ultrasound guidance. In the present situation of an ongoing obesity epidemic where the use of regional anesthesia is advantageous primarily for pulmonary reasons, such improvements would be welcomed, especially when combined with nerve-specific imaging such as Impedance Neurography.

EXTRACTING TISSUE TIME CONSTANTS

To selectively stimulate neuronal subpopulations, it would be advantageous to ascertain time constants of nerves for nerve stimulation applications, e.g., neuromodulation, but is that possible? This ability would be especially useful when disease states or anatomy leads to different electrical behavior of the neuronal cell membrane, e.g., diabetic neuropathy, where it has long been recognized that effective nerve stimulation requires higher current outputs. This is a reflection of the change in the p-RC charging characteristics of the diabetic nerve.

Recall that the electrical decay curve following the termination of a voltage or current pulse is exponential per Eq. (3.4). In fact, there are multiple RC time constants involved due to the various components of the tissue electrical path, which includes more than just nerve, and the voltage decay curve corresponds to Eq. (3.15):

$$V = \sum \left(C_0 e^{-t/\tau_0} + C_1 e^{-t/\tau_1} + C_2 e^{-t/\tau_2} + \cdots + C_n e^{-t/\tau_N} \right) \qquad (3.15)$$

where τ_{0-N} are the component time constants and C_{0-n} are constants. This can be illustrated in Fig. 3.37.

Fig. 3.37 shows the voltage response to a current pulse of 0.489 mA delivered at a depth of 1 cm just overlying the course of the saphenous nerve in the calf. Inspection of the decay portion of the voltage response, following termination of the current pulse reveals the exponential decline of developed voltage toward zero. However, by graphing the natural log of the decay curve, Fig. 3.38, it's obvious that a straight-line result is not obtained, i.e., it's not a single-term exponential.

The component time constants may be derived in the following fashion using a well-known technique for separation of radioactive decay half-lives from mixtures of radioisotopes: logarithmic peeling.[12]

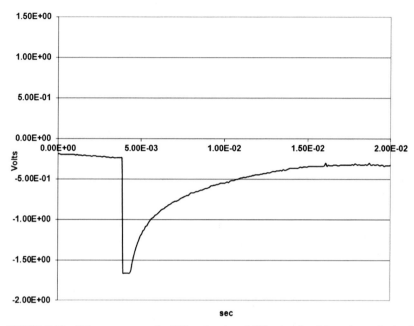

FIGURE 3.37   Voltage response of a 200 μs duration, 0.489 mA pulse delivered at a depth of 1 cm over the course of the saphenous nerve.

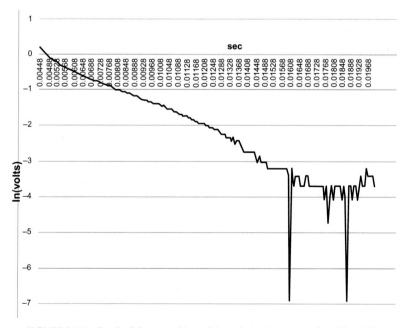

FIGURE 3.38   Graph of the natural log of the voltage decay curve from Fig. 3.37.

If one of the transients in Eq. (3.15) has a long duration time constant relative to the other terms in the summation, it will dominate the characteristic waveform of the voltage curve and account for the late portion of Fig. 3.38, labeled as V[tail] (following the nomenclature of Rall) in Eq. (3.16):

$$V[tail] = C_0 e^{-t/\tau_0} \qquad (3.16)$$

And taking the natural log of V[tail] gives Eq. (3.17):

$$\ln(V[tail]) = -t/\tau_0 + constant \qquad (3.17)$$

Rearrangement of Eq. (3.17) gives Eq. (3.18):

$$\tau_0 = -1/slope \ of \ \ln(V[tail]) \qquad (3.18)$$

where slope means the change of ln(V[tail]) with respect to time.

Using this information, a theoretical curve may be constructed to the earlier time points and the calculated values subtracted from the observed measurements to "peel" away the first transient, shown in Eq. (3.19):

$$V - V[tail] = V[peeled] \qquad (3.19)$$

Since τ_0 is large relative to τ_1, the same process is repeated to peel the second transient. Calculating the slope and intercept between times 0.01216 s and 0.02 s on Fig. 3.38 gives Eq. (3.20):

$$\ln(V) = -254.046t - 0.3071 \qquad (3.20)$$

shown graphically in Fig. 3.39.

This result was used to calculate the decay curve of the associated time constant as follows in Eq. (3.21):

$$V[tail] = e^{(-254.046t - 0.3071)} \qquad (3.21)$$

This process is shown graphically in Fig. 3.40.

When V[tail] 1 is subtracted from V, V[peel] 1 is obtained. Once again taking the natural log of V[peel] 1, the graph in Fig. 3.41 is obtained.

The slope and intercept of the straight portion of ln(V[peel] 1) is shown graphically in Fig. 3.42.

The above process is repeated iteratively until Figs. 3.43–3.45 are obtained.

Note that in Fig. 3.43 two exponential functions have been extracted, one having the parameters in Eq. (3.20) the other (Eq. 3.22) conforming to the dashed line in Fig. 3.42:

$$V[tail] 2 = e^{(-189.837t - 1.8314)} \qquad (3.22)$$

The resulting curve, V[peel] 2 is shown to be the result of a single exponential in Fig. 3.44 where V[tail] 3 closely overlaps V[peel] 2, and has the parameters of Eq. (3.23):

$$V[tail] 3 = e^{(-2544.07t - 0.9931)} \qquad (3.23)$$

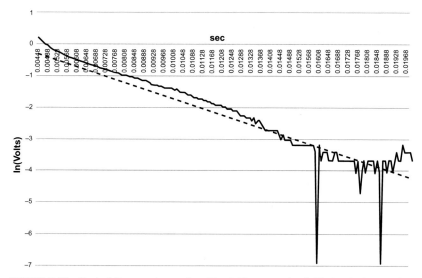

FIGURE 3.39 *Dashed line* superimposed on Fig. 3.38 representing ln(V) for the longest time constant.

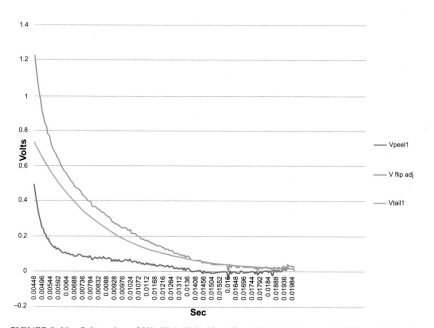

FIGURE 3.40 Subtraction of V[tail] 1 (light blue) from V (dark blue) to yield V[peel] 1 (red).

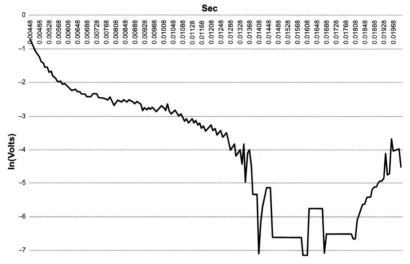

FIGURE 3.41 Natural log graph of V[peel] 1.

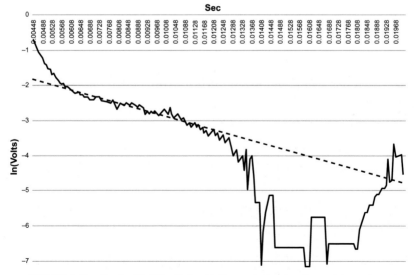

FIGURE 3.42 Graph of ln(V[peel] 1) with line approximating the straight portion.

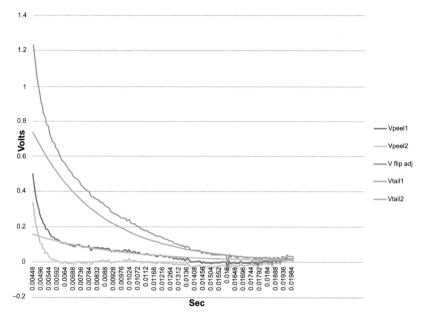

FIGURE 3.43 Curves for V[tail] 1 and V[tail] 2 subtracted from V leaving V[peel] 2.

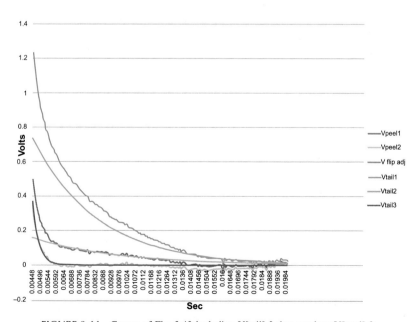

FIGURE 3.44 Curves of Fig. 3.43 including V[tail] 3 that overlaps V[peel] 2.

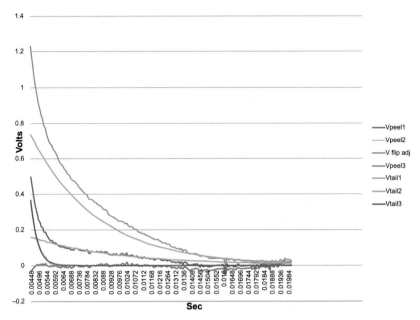

FIGURE 3.45 All three exponentials, V[tail] 1, V[tail] 2, and V[tail] 3, subtracted from curve V yield a result of zero (V[peel] 3).

Finally, in Fig. 3.45, the three derived exponentials have been subtracted from curve V and the result is zero.

The voltage decay data V are shown to consist of at least three exponentials with time constants of the form in Eq. (3.18), giving values of $\tau_0 = 3.936$ ms, $\tau_1 = 5.267$ ms, and $\tau_2 = 0.393$ ms.

The derived time constants show that τ_0 and τ_1 are both in the range of those found for mammalian spinomotoneurons according to Rall.[2] Studies conducted over several peripheral nerves have shown that it is the second time constant, in this case τ_1, which most closely correlates with the right-angle relationship to the position of nerves. The second time constant has been found to decrease with lateral distance from the position of the peripheral nerve, whereas τ_0 and τ_2 do not change with lateral distance from the nerve.

This kind of analysis of voltage decay data following single current or voltage pulses in tissue can be of advantage for determining best stimulation parameters, particularly pulse width, in neuromodulation applications. Also, the ability to extract component time constants may provide a means to detect skin breaks. If the keratinized layer of skin is disrupted, e.g., by a scratch, an impedance determination from that scratched area will be low. This is a source of error for Impedance Neurography and a means to determine if a low impedance site was related to nerve or indicated a skin break would be useful. Rapid determination of the component time constants may provide an answer. Further work will be required to ascertain how to best derive the appropriate time constant, but this technique provides information that it is possible to do so.

REFERENCES

1. Cole KS. *Membranes, ions and impulses. Vol biophysics series*, vol. 1. Berkely, Los Angeles, London: University of California Press; 1972.
2. Rall W. Core conductor theory and cable properties of neurons. In: Kandel E, editor. *Handbook of physiology Section 1: the nervous system, Volume 1. Cellular biology of neurons, part 1*. Bethesda: American Physiological Society; 1977.
3. Hodgkin AL, Huxley AF. A quantitative desription of membrane current and its application to conduction and excitation in nerve. *J Physiol* 1952;**117**:500−44.
4. Mauro A. Anomalous impedance, a phenomenologic property of time-variant resistance. *Biophysical J* 1961;**1**:353−72.
5. Raymond SCD. *Inventor. apparatus and method for locating a nerve*. July 7, 1998. US Patent 5,775,331.
6. Cooper M. Membrane potential perturbations induced in tissue cells by pulsed electric fields. *Bioelectromagnetics* 1995;**16**:255−62.
7. Tai C, Guo D, Wang J, Roppolo J, de Groat W. Mechanism of conduction block in amphibian myelinated axon induced by biphasic electrical current at ultra-high frequency. *J Comput Neurosci* November 2011;**31**(3):615−23.
8. Cooper M. Gap junctions increase the sensitivity of tissue cells to exogenous electrical fields. *J Theor Biol* 1984;**111**:123−30.
9. Cooper M, Miller J, Fraser S. Electrophoretic repatterning of charged cytoplasmic molecules within tissues coupled by gap junctions by externally applied electric fields. *Dev Biol* 1989:179−88.
10. Sabah N, Leibovic K. Subthreshold oscillatory responses of the Hodgkin-Huxley cable model for the giant squid axon. *Biophysical J* 1969;**9**:1206−22.
11. Ranck J. Which elements are excited in electrical stimulation of mammalian central nervous system: a review. *Brain Res* 1975;**98**:417−40.
12. Rall W. Membrane potential transients and membrane time constant of motoneurons. *Exp Neurol* 1960;**2**:503−32.

Chapter 4

Anisotropicity

Early in the process of investigating Impedance Neurography I approached a neuroscientist who worked with anesthetized cats and rabbits doing brain-related research. I proposed to him that while he was working on the head end of an anesthetized rabbit, if he had no objections, I could be working on the hind end, the region of the sciatic nerve. I wanted to demonstrate the Transcutaneous Electrical Nerve Stimulation (TENS) effect as a way of forging a relationship for further study of the effect seen during my clinical observations. It was an interesting endeavor for a couple of reasons, one related to neuroscience and the other related to olfactory observations.

To effectively couple the examiner's finger and the subject's skin, conductive gel must be placed between the finger and the skin. As you can imagine, this is complicated if the subject is furry, and rabbit legs are definitely furry. I knew from my work with human patients that any nicks or cuts in the skin would impair my ability to determine changes in the electric shock sensation in my finger since interrupting the skin surface created a low impedance path down which the current could flow. That knowledge informed us that we could not simply shave the rabbit's leg since small cuts would invariably occur.

Enter Nair…

What a great idea! Nair is a depilating preparation for removing hair without using sharp objects. After sufficient contact time, one can simply wipe the hair, or fur in this case, away. Well, there are some things the package inserts do not tell one unfamiliar with using this material. The worst is…it stinks; it really stinks. Nair works by breaking disulfide bonds in hair and some of that sulfur is reduced to hydrogen sulfide gas of Yellowstone Park hot spring fame, i.e., the "rotten egg" smell. The effect is not immediate, so all the Nair had been applied before we began to notice the stench that was related to the amount of hair involved and, again, a rabbit leg is very hairy. It was effective, however, and the hair could be wiped away in a hair-goop mixture that required a surprising number of paper towels to clean up. As you might imagine, we only did this once.

The second reason that this experiment was interesting was the result. Using the TENS technique with my finger as the probe, I was able to find the path of the sciatic nerve in the anesthetized rabbit's thigh and mark its course on the skin surface with a skin scribe (a gentian violet pen for drawing on the

Finding the Nerve. http://dx.doi.org/10.1016/B978-0-12-814176-2.00004-6

skin surface). Then, using a nerve stimulator, I advanced a stimulating needle through the skin at a point on the line I had drawn and on a normal to the skin surface until the rabbit's foot began to twitch in time to the stimulator beep. At that point a stainless-steel wire was advanced through the needle to its tip and the needle withdrawn leaving the wire in place. My neuroscience colleague then dissected on the usual approach to expose the sciatic nerve, which was different than the path down which I had inserted the needle that was indicated by my measurements and skin marking for the right-angle relationship. He was peering in the hole of his dissection when, at one point, he began to laugh. On my asking what was so funny, he replied, "You stuck the wire right through the nerve." Needless to say, he was convinced that the effect was real. Before the experiment, I believe he was quite skeptical even though he was too polite to say so.

THE IMPORTANCE OF ASSUMPTIONS

One of the tacit assumptions underlying Impedance Tomography and other medical techniques involving externally applied electric fields is that tissue can be described as a bulk conductor, albeit with imbedded electrical non-homogeneities represented by different tissues. Resistivity prospecting for ore bodies is successfully conducted with similar assumptions where current flow will distribute differently in areas (rocks) of different conductance character-istics.[1] In fact, resistivity prospecting provided a great many pertinent facts informing some of my nascent studies. But, early on in tissue impedance investigations I noted the confusing finding that the underlying nerves were invariably found at a right angle to the skin surface. This was a striking observation without a ready explanation; however, it held up to investigations such as the rabbit study. In fact, the relationship was consistent enough that if a nerve stimulator was not used to determine how close the needle tip was to the nerve, skewering the nerve was a definite risk, as learned from the dissection work.

We embarked on some studies to try and model the right-angle effect and initially thought using nerve suspended in a bath would be a promising avenue to pursue. The question of what we could use as a source of animal nerve presented an early hurdle and, among other options, we seriously debated obtaining mammalian nerve from slaughterhouses. This seemed a more difficult course of action since I had no idea how close any such facilities might be and was not particularly interested in finding out. Fortunately, there was a convenient alternative.

The squid giant axon model has been well established in research since the 1930s, but it turns out that giant axons are not limited to that species. They occur in some insects, notably grasshoppers, and crustaceans. Insect material was simply too short to be of any use, but live lobsters are kept in some su-permarkets and their ventral nerve was a good option. Plus, using lobsters as sources for study had the added advantage that if carefully handled, we could enjoy a nice meal afterward.

The usual way lobsters are dispatched for culinary purposes is to toss them in a pot of boiling water. I found this a distasteful approach as it seems a bit cruel, plus it would cook the nerve along with the rest of the creature. A better, quicker, and more humane approach was to pith the animals. This involves a very quick thrust of a sharp knife, and the process is over in a fraction of a second, but the ventral nerve remains functional even after the lobster as a whole is not. The nerve can be painlessly dissected and then mounted in a bath of lobster Ringer's solution for additional study. A nuance to working with the lobster material was, in contrast to the squid nerve, there were multiple, segmental branches coming off the lobster ventral nerve that had to be ligated and cut during the dissection. If this technique was not followed, the axoplasm would simply leak out of all the various cut branches rendering the nerve nonfunctional very quickly once mounted in the bath.

Our goal was to determine whether the nerve was detectable in the bath of lobster Ringer's solution when the sampling electrodes were not in contact with the nerve itself. To accomplish this, sampling electrodes had to be constructed from material that would not react with the salts in the lobster Ringer's solution. We tried a couple of options.

One electrode system we evaluated consisted of dissecting the Ag:AgCl-coated, plastic buttons out of ECG electrodes and mounting them in a row on a piece of polystyrene foam that would float on top of the Ringer's solution. This arrangement worked quite well, but we were unable to show any ability to detect the nerve suspended in the bath. This was true even when we verified that action potential generation was possible in the nerve preparation.

The lack of success using Ag:AgCl electrodes prompted a search of another system and I found that a 2 mm pencil lead fit very snuggly in a 20G IV catheter. It was possible to make these constructs with any useful length of pencil lead exposed at the tip and attach an electrode to the other end of the lead that protruded from the hub of the IV catheter via an alligator clip. The pencil leads did not present an unusually high resistance, but the results when tested in the bath containing the lobster ventral nerve were the same. No ability to detect the suspended nerve was noted.

The lobster nerve model was discarded after these repeated failures with some reluctance, given we had several good meals after the experiments. I decided to try the most conductive thing I could think of and placed an 18G copper wire in the bath. Once again my efforts were notable only for their lack of success. There was no evidence that the wire could be detected.

These results were confusing since it was clearly possible to detect a nerve in a living mammal, but not possible, using essentially the same equipment, with an isolated nerve or wire preparation. I even went to the extreme of placing a sterile epidural catheter guide wire under my skin and tried to detect it. No luck. As I thought about the situation, there was one difference between the two setups that was pretty obvious. In the living situation, small nerves branch off the larger, underlying nerve and travel up to the dermis. Could these smaller fibers be playing a role in our detection results?

I constructed a model using an 18G copper wire with four 0.025" diameter, uncoated copper wires wound around the larger wire leaving two, long tails of each fine wire piece for a total of eight tails. Then, I took a block of open-cell foam sponge that was cut to the dimensions of the container we used for our nerve preparation bath and placed the long, 18G wire on its surface. Each, long tail of the fine wires wound around the larger wire was then threaded through a needle and inserted through the foam to the opposite side on various trajectories with insertion angles of between ~ 10 and ~ 45 degrees. On finishing the labor-intensive part of threading all those fine wires, when turned over, the foam appeared to have grown some of coppery hair. Gently depressing the foam, the fine copper wires were trimmed so that the cut end was around 2–3 mm beneath the sponge surface when the foam was allowed to spring back into shape. This was done at two, separate places on the 18G copper wire, roughly 6 inches apart. One was for the sampling electrode array, the other for the return electrode.

A second sponge, cut to the same dimensions as the first, was placed in the bottom of an acrylic container designed to hold the sponges in a bath solution. The wired foam was then placed in the bath container on top of the first, unwired sponge and the container was filled with saline solution to just saturate the surface of the foam. A cross section of the setup is shown in Fig. 4.1.

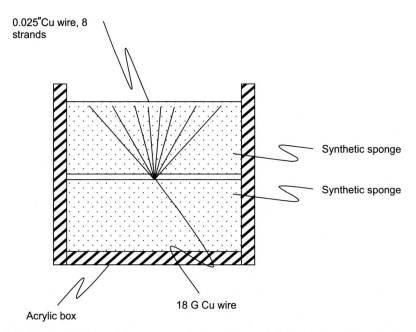

FIGURE 4.1 Depiction of the study setup for detecting an underlying wire in a saline bath using 0.025" diameter copper wires connected to an 18G copper wire. The fine wires have been trimmed so their cut ends lie just deep to the surface of the saline bath in the substance of saturated, open-cell sponge material. *(Data courtesy of Nervonix, Inc. Image construction by Philip C. Cory, M.D.)*

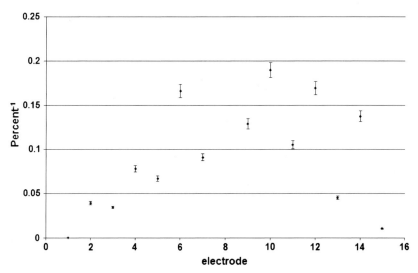

FIGURE 4.2 Voltage differences, graphed as reciprocal percentage differences, from sampling sites in line with the line of cut ends of the 0.025" diameter copper wires connected to an underlying 18G wire as shown in Fig. 4.1. *(Data courtesy of Nervonix, Inc. Image construction by Philip C. Cory, M.D.)*

Using the original equipment described in the Introduction, connected to a flex circuit electrode array having two, offset rows of seven electrodes each, the array was placed along the line of trimmed, fine copper wires without any care to having individual electrodes placed in any particular relationship to the wires themselves. Sampling was performed and the results were rather dramatic.

The order of electrodes shown in Fig. 4.2 was the distal, left electrode was designated "1." The proximal left electrode was designated "2" and the distal second-most leftward electrode designated "3," and so forth for a total of 14 electrodes. The electrodes in the proximal row were closer to the position of the fine copper wires; hence, their applied voltages were lower than the corresponding electrode in the distal row. This gives the staggered appearance on the graph in Fig. 4.2. The results are depicted as reciprocal percentage differences giving the largest numeric value to the lowest voltage electrode. The device used functioned in a constant current mode, so the applied voltage tracked the impedance, i.e., the lower the applied voltage, the lower the impedance.

I had found a model that mimicked the *in vivo* results from the mammalian subjects (including human) that had been studied. Though it took several more years to understand what was taking place with this rather simple model, it was the first demonstration of the three-element model discussed in Chapter 3. In this case the three elements were a thin layer of saline solution, in series with a fine copper wire, in series with a large copper wire. The time-variant conductances that I have discussed in the three-element nerve model were represented by the electrode contact impedance, the surface capacitance characteristics of the copper wires, and the inductance of the copper wires. These well-known characteristics of the wires behaved in a similar fashion to the time-variant resistances of the voltage-gated channels of the nerves.

The above work from the isolated nerve and wire studies, as well as from the frequency studies reported in Chapter 3, helped to reveal the short answer to this confusing finding of a right-angle relationship. Simply stated, living tissue is not just a bulk conductor with variable conductance regions. Such a bulk conductor, even containing variable electrical conductance regions, demonstrates no directional preference for current flow and is described as nonhomogeneous (variable conductance regions) and isotropic (no directional specificity to current flow). However, we have found that living tissue displays marked conductance facilitation along specific tracks. This characteristic of facilitated current conduction along specific paths found in living tissue means tissue is *an*isotropic to an applied current and that anisotropicity can be enhanced using a sampling system with appropriately designed electrical parameters. This finding does not apply to dead tissue that lacks the ion channel time-variant conductances, and all studies in the literature evaluating the electrical parameters of tissue use cadaveric tissue. It may be freshly cadaveric, but still cadaveric since designing a study to test the electrical parameters of a block of specific living tissue such as muscle, bone, liver, etc., is a very difficult system to set up. So, technologies such as Impedance Tomography that depend on applying electric fields to tissue point to some extremely well-done studies from cadaveric tissue as the basis for their application. Typical is the work of Gabriel et al.[2]

A REVELATION

I found the results discussed above interesting but had a problem determining the underlying mechanism accounting for the observed anisotropicity in living tissue: the right-angle relationship. The wire studies were performed in 1998, but the "light bulb" episode occurred when I saw the following image from *Pain*, 98(1−2), 2002, front cover.

Fig. 4.3 depicts a photomicrograph of skin stained for neurons, the small, green fibers oriented at right angles to the skin surface.

These are the only structures I know of that course at right angles to the skin surface in high density. But could nerves explain the anisotropicity of living tissue? If this were true, it would help to verify the observations from the wire in saline studies. The answer depends on the nerve cells having intact and functional cell membranes, a situation that only pertains in the living state. The intact membrane provides the high capacitance necessary for the low

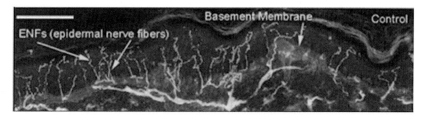

FIGURE 4.3 Photomicrograph of skin stained for nerve fibers (green).

impedance represented by nerve cells, so nerves could clearly function as the path of least resistance to current flow based on their electrical cable conduction characteristics alone. Such cable core conduction properties of nerves described by electrotonics are perhaps best discussed in the chapter by Rall, but nerves have other, additional factors that facilitate their conductance of current.

Cooper et al. discusses the ability of long, uninterrupted cells, or chains of cells connected by gap junctions, to function as facilitated conduction paths.[3,4] The reason for this effect is the absence of membrane barriers to current flow inside the cells. In normal tissue, current crosses two membrane barriers for each cell in its path: one on entering the cell and another on exiting the cell. Both membranes constitute impedance barriers or hurdles over which the current must leap. In a neuron, the distance between impedance hurdles can be a meter or more in length. So, when a minimal current is applied to a subject between an electrode on the foot and another on the forehead, such as for Impedance Tomography, rather than distributing as though flowing through a bulk conductor, a majority of the current preferentially distributes largely into the nerves of the toe, travels up to the spinal cord, continues up the spinal cord to the head and neck, and travels out the nerves to the skin where the return electrode resides, taking advantage of the cable/core conductance properties of neurons. The lack of recognition of this property of living tissue current flow likely explains the difference between the theoretical precision of Impedance Tomography versus its practical application, i.e., it's never lived up to its promise. Many very sophisticated back-projection algorithms constructed to try and solve the so-called inverse problem and predict the structure of underlying tissue based on surface voltage distributions have failed to provide the resolution desired. This is largely because the basic, underlying concept of electrical field distribution used by Impedance Tomography theory in tissue is incomplete.

An additional, neuron-specific, factor leading to current flow anisotropic behavior of nerves is the anomalous impedance described by Cole and further elucidated by Mauro, and Sabah and Leibovic as discussed in Chapter 3.[5–7] Cole's anomalous impedance is a function of the variable resistances that voltage-gated sodium and potassium channels demonstrate to applied current. If these channels are biased with a continuous current flow through the cell membrane, and a periodic waveform (e.g., a sinusoid) is applied to that continuous current flow, the resistances of the channels display exponential increases and decreases over time depending on the direction of current flow and the channel in question (inwardly facing, e.g., sodium channel, or outwardly facing, e.g., potassium channel). These exponential resistance changes with time mimic those of capacitive and inductive reactances and show predictable relationships to the frequency of the applied periodic waveform. Depending on whether the capacitive reactance-like or inductive reactance-like effect predominates, impedance can fall or rise with a particular frequency. At frequencies less than around 3 kHz, impedances are low along the course of neurons compared to tissues with low nerve densities, whereas at frequencies greater than 6 kHz, impedances rise and impedance differences between nerve-rich and nerve-poor tissue are lost. Since most studies of tissue electrical properties are performed with frequencies of 10^5-10^9 Hz, and with equipment using good electrical engineering design as discussed earlier, it is clear why this anisotropic behavior has not been observed.

NERVES ALONE ARE IMAGED

In the mid-1990s, I presented a couple of posters at a meeting documenting some of our early findings. At that time, I had absolutely no idea regarding the underlying mechanisms involved, but thought the results interesting enough to present to the clinical and scientific communities. Many meeting participants strolled past my posters without comment, though a few did ask some questions. Consequently, I was very pleased when a physician, well known in the area of muscle disorders, stopped by with a string of admirers in tow. He listened for about 30 seconds before telling me that what we had observed could not happen because everything deep to the skin was essentially a salt bath and would not behave the way our carefully collected data showed. His diatribe went on for about 10 min with all the admiring hoard dutifully listening and nodding their heads. He closed by smiling, giving me a paternalistic pat on the shoulder and headed off to his next destination with his admiring throng following in step.

This episode was publicly embarrassing, but felt wrong, and not just for the embarrassment. In retrospect, it was a good demonstration of how assumptions can lead us astray. Something about his assessment just did not ring true and his pronouncements about how the world really worked made me more interested in determining how it was that he was mistaken, especially in light of how readily he dismissed carefully collected data that did not comport with his worldview.

Two years later, I attended another meeting, this time as an exhibitor. We had completed work on our second prototype device that looked a lot like a channel changer with a dinner fork attached to it. A third, more sophisticated prototype was under development and it seemed a good time to assess commercial interest. Besides, the meeting was in Vienna and would have been difficult to turn down. The same individual stopped by our booth, took a look at what we were demonstrating, and became a convert on the spot. He went from being a major detractor to an enthusiastic supporter in the space of a few minutes. What changed his mind? In a word, images. At the first meeting, my posters simply presented data and showed our experimental setup. By the second meeting, we had progressed to the point that it was possible to create 2-D images on the skin surface of the underlying neuroanatomy and he could "see" what we were talking about. What was really interesting was that on the last day of the meeting, another researcher/participant stopped, looked at our work, and announced, "You can't do that." He then turned on his heel and left, having provided us with his terse, revealed wisdom.

I find it constantly amazing how people from all walks of life can look at obvious results and deny their reality without delving into the background underpinning the observations in any way. Admittedly, many do not have the grounding in appropriate areas of study to enable knowledgeable scrutiny of the underlying facts and their denial reflects a basic insecurity about the nature of existence. Others with the skill set required to accurately evaluate data may choose not to because the data do not comport with their view of how things really are. Both approaches make life much more interesting for the rest of us. Thomas S. Kuhn touches on this in his *The Structure of Scientific Revolutions* (University of Chicago Press, 1962), which is a book well worth reading.

The doubting initially expressed by the eventual convert as well as the terse "expert" goaded me on and it was because of their points of view that I was led to discover the works of Ken Cole and Wil Rall to which I have referred many times in this work, wherein a more complete understanding of tissue electrical responses was revealed.

I have had the privilege of corresponding with or meeting a few of the great names in neuroscience. Patrick Wall is recognized in the dedication for his encouragement. Fortunately, I have had a long-standing friendship with one of Wil Rall's graduate students and because of that got to meet Dr. Rall himself, finding him to be a very gracious and humble man, whose interests had moved from neuroscience to sculpture by the time I met him. Dr. Rall is one of the founders of the modern discipline of computational neuroscience, bringing his background in physics and math to bear on knotty problems of neuronal transmission and the excitation of nerve membranes. He was attending Yale during WWII, but rather than being drafted as were many of his classmates, because of his studies in science, he was allowed to finish his degree to become part of the scientific manpower pool. On graduating, he joined the Manhattan Project through the Physics Department at the University of Chicago along with Ken Cole. There he helped develop an improved version of the mass spectrograph needed in the production of plutonium. He decided to switch to biophysics after the War and worked with Ken Cole at Woods Hole Institute and the University of Chicago where he mingled with Alan Hodgkin, Enrico

Fermi, Sewall Wright, and others. It must have been a heady time. He eventually sojourned to New Zealand to work with J.C. Eccles, himself a former student none other than Sir Charles Scott Sherrington. This was, indeed, a formidable lineage in neuroscience.

What I learned from our work, and reading Cole and Rall, was that far from being a simple salt bath, the behavior of tissue subjected to electric fields is complex and held some surprises for me to discover. It was also interesting that none of this area of research had ever been presented in my medical school training. Perhaps, in light of my training experience, it was not surprising that other physicians would know nothing about it, either. It took another decade to ferret out the particulars of how tissue works as an electrical system, but even after obtaining a good understanding of the current state of the research and thoughts regarding equivalent electrical circuit models of tissue, something was still amiss.

It was not until data were obtained, some of which were presented in Fig. 3.3, showing the parallel resonance phenomenon that everything began to fall into place. That occurred in 2012, over 20 years after the first observations with the TENS technique were made. All of this brings to mind my favorite Josh Billings paraphrase, sometimes misattributed to Mark Twain:

> *"We don't get into trouble nearly as much for what we don't know as we do for what we know that just ain't so."*

Factors discussed above, the RLC behavior of neurons, the transmission properties of long, uninterrupted tubular structures, and the high density of voltage-gated channels in neurons explain why they alone are imaged with Impedance Neurography. Other tissue types lack one or more of the critical components that make neurons unique electrical transmission systems.

Blood and lymphatic vessels are also long, tubular structures, but they lack the high density of voltage-gated channels and have a wealth of cell membranes in their interior to act as impedance barriers. Though it is tempting to assume that current can circumnavigate cells in vessels or tissue flowing entirely through the serum or interstitial fluid, which I was actually taught in medical school, current does not just flow around objects suspended in fluid. This was shown by Cole in impedance studies of milk (fat globules), suspensions of *Arbacia* eggs, and packed red blood cells,[5] and is a basic understanding in applications such as Impedance Tomography.

Remember the T-Scan 2000 ED from earlier? The premise behind that device was based on the observations that some cancer cells display increased electrical capacitance and that those cells may be detectable through the use of skin surface impedance determinations at multiple frequencies. From the discussion above, the reason this is not possible can be determined.

Although some cancer cells display high capacitance, they are neither long transmission structures nor do they behave as RLC circuits under appropriate conditions, explaining the inability of impedance determinations to detect clumps of cancerous cells using a two-electrode system as used in Impedance

Neurography and was attempted with the T-Scan 2000 ED. Analogously to resistivity prospecting and Impedance Tomography, one might expect that a three-electrode system could potentially "see" these clumps of cells due to their constituting an electrical nonhomogeneity in the tissue. There is another approach to consider, however.

Malignant clumps of cells may displace nerves, leading to a means of detecting tumors through "Stealth" technology: not seeing something where one expects to see it. Stealth approaches are one of the intriguing possibilities raised by Impedance Neurography. This is particularly true if one considers that there exists a background of nerve-related tissue anisotropicity related to the myriad of axons traveling through tissue. This is like the background of light impinging on a camera where if the aperture is opened completely, a white-out occurs. It is only with restricting the amount of light impinging on the camera through partial closure of the aperture that individual objects become discernable. Similarly, the background impedance change related to the myriad of axons in tissue may be changed by the presence of objects, e.g., a tumor mass, which is not innervated normally and displays an abnormal background distribution of axons. It's an intriguing possibility.

THE HIERARCHY OF NERVE STRUCTURES IMAGED

This lowered impedance associated with nerves also explains the hierarchy of nerve structures visualized by Impedance Neurography that I brought up in the Introduction. Importantly, Impedance Neurography is nerve specific in that it only images nerve tissue, demonstrating no ability to image muscle, vessel, or internal organs as Impedance Tomography techniques can to some degree. Normal nerve is demonstrated by its low impedance when compared to surrounding tissue. Even lower impedances are seen from some other nerve structures such as nerve branch points that often correspond to classic acupuncture sites or myofascial trigger points. Entrapments and nerve contusions show lower impedances yet, probably related to variable degrees of demyelination associated with those conditions, while neuromas demonstrate the lowest impedance measurements. This hierarchy corresponds with the amount of exposed (unmyelinated) neuronal cell membrane in the image field. Impedance Neurography appears to be a method of visualizing exposed neuronal cell membrane and assessing its electrical properties. Notably, neurons determine where to place voltage-gated channels in the membrane by lack of myelination. Therefore, with increasing density of unmyelinated neuronal cell membrane in the imaging field, the determined impedance will decrease. Of the above structures, neuromas have the most exposed neuronal cell membrane and the highest densities of sodium and potassium channels of any known tissue.[8,9] This corresponds to neuromas showing the lowest impedances with the application of Impedance Neurography.

Branch points of nerves are an interesting case. Why do they display lower impedances than normal nerve? After all, branching is a function of normal nerve. Consider that at a branch point, the cell membrane to volume ratio approximately doubles, and suddenly there is a much higher density of neuronal cell membrane. This observation fits well as a mechanism behind the visualization of branch points as lower impedance sites than normal nerve. Branch points may be vertical, as in the example from Chapter 1 where the underlying anatomy of a myofascial trigger point was elucidated, or they may be horizontal. Some subcutaneous acupuncture points, unrelated to muscle, are such horizontal structures. Due to some nerve branch points being very consistent from person to person, it's possible to construct maps of their positions on the human body. This is the source of the acupuncture diagrams used in acupuncture therapy. The meridians depicted on the skin surface, unfortunately do not exist. However, there is the possibility that interconnections at a higher level in the CNS, e.g., the brain, may account for some of the apparent relationships. Mostly, the lines drawn on the skin surface are an example of how our brains connect the dots, just like in constellations in the sky where, since ancient times, we have seen all sorts of connections. The important thing is that we have good, reproducible data that acupuncture is effective treatment for some people. It is clearly associated with an enkephalinergic effect in the spinal cord and shares that characteristic with low frequency TENS therapy. In both situations, if an individual who has shown pain reduction with acupuncture or TENS therapy is given naloxone (an opioid receptor—blocking agent) pretreatment, he or she no longer derives pain relief from the therapies.

It also appears that the state of activation of nerves is appreciable by Impedance Neurography. Recall in the Introduction, I discussed a woman with low back pain and showed an impedance neurograph of her sacral region. In addition to the parallel columns of impedance minima corresponding to the nerve-filled sacral neuroforamina, her imaging also displayed two impedance minima lateral to the left neuroforamina that were found to correlate with the course of articular branches of the sacral roots supplying sensation to the sacroiliac joint. There were no corresponding elements seen on the right side, though articular branches had to be present there, too. It appeared that because the left-sided articular branches were actively relaying nociceptive information, they were revealed as low impedance sites. Since these nerves were firing repetitively, the equilibrium of sodium and potassium channel states (open vs. closed vs. inactive) were different on the left compared to the right. This was the first evidence that Impedance Neurography reveals information about the activity of peripheral nerves and has been observed in other situations as well. Actively trying to stimulate nerves with a nerve stimulator during Impedance Neurography has yielded mixed results, usually demonstrating higher than expected impedance during active stimulation. The explanation may relate to the fiber class that is being stimulated with electrical nerve stimulation. A discussion relative to this fiber class consideration is presented in the next chapter, though for different reasons. The bottom line is that this is an area ripe for additional investigation.

REFERENCES

1. van Nostrand R, Cook K. *Interpretation of resistivity data*. Washington, D.C.: United States Government Printing Office; 1966.
2. Gabriel S, Lau R, Gabriel C. The dielectric properties of biological tissues: I. Literature survey. *Phys Med Biol* 1966;**41**(11):2231−49.
3. Cooper M. Gap junctions increase the sensitivity of tissue cells to exogenous electrical fields. *J Theor Biol* 1984;**111**:123−30.
4. Cooper M, Miller J, Fraser S. Electrophoretic repatterning of charged cytoplasmic molecules within tissues coupled by gap junctions by externally applied electric fields. *Dev Biol* 1989:179−88.
5. Cole KS. Membranes, ions and impulses. In: *Vol biophysics series*, vol. 1. Berkely, Los Angeles, London: University of California Press; 1972.
6. Mauro A. Anomalous impedance, a phenomenologic property of time-variant resistance. *Biophysical J* 1961;**1**:353−72.
7. Sabah N, Leibovic K. Subthreshold oscillatory responses of the Hodgkin-Huxley cable model for the giant squid axon. *Biophysical J* 1969;**9**:1206−22.
8. England J, Happel L, Kline D, et al. Sodium channel accumulation in humnas with painful neuromas. *Neurology* July 1996;**47**(1):272−6.
9. England J, Happel L, Thouron C, Kline D. Abnormal distributions of potassium channels in human neuromas. *Neurosci Lett* October 1998;**255**(1):37−40.

Chapter 5

Depth Determination of Peripheral Nerves Using Impedance Neurography

ELECTRODE SEPARATION DISTANCE CONSIDERATIONS

Very early in Impedance Neurography work, the observation was made that as the skin surface electrodes were separated to greater distances over the course of a nerve, the resistance fell. This is shown by data in Fig. 5.1 from an early study of the lateral digital nerve of the thumb.

For a time, the observed fall in resistance was confusing, particularly when determined over a digit that did not appreciably change its diameter in the

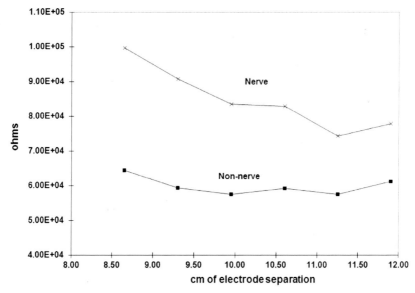

FIGURE 5.1   Resistance versus electrode separation distance measured over the dorsolateral digital nerve of the thumb.

Finding the Nerve. http://dx.doi.org/10.1016/B978-0-12-814176-2.00005-8

study region (no increase in electrical path cross section). For some, the concept of electrical resistance of tissue as an electrolyte may not be entirely clear and a discussion of considerations related to tissue electrical properties is in order.

Biological tissue, though complex in its organization, from an electrical point of view is largely a solution of ions in water with some extremely important differences. In other words, tissue is not just a salt bath! When current flows through an ionic medium, such as tissue, the movement of charged entities is not unidirectional as is electron flow in metallic media such as wires. If a battery is connected across the length of a wire, electrons move away from the negative potential toward the positive potential, and it is those electrons moving as charge per unit time that constitute current flow. The situation is different in ionic media where, due to the conservation of charge, an equal number of negatively charged ions and positively charged ions exist. It is also important that these ions are floating around in a medium, water, that is termed a dipole. And just what, exactly, is a dipole?

Remember the molecular structure of water: H_2O. Two hydrogen atoms are bound to one oxygen atom (Fig. 5.2).

Since the oxygen atom needs two electrons to fill its outermost electron shell, the electrons from the hydrogen atoms are attracted toward the oxygen, leaving a relative positive potential in the vicinity of each hydrogen atom. Similarly, the region surrounding the oxygen atom has a relative negative potential. These separated potentials constitute a molecular dipole (Fig. 5.3, two regions of opposite charge—dipolarity).

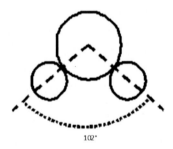

FIGURE 5.2 Diagram of a water molecule with two hydrogen atoms attached to one oxygen atom at an interatomic angle of 102 degrees.

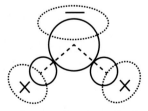

FIGURE 5.3   Diagram of a water molecule as a dipole.

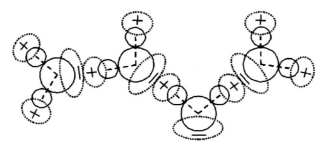

FIGURE 5.4 Depiction of water molecules associating based on the attraction between dipoles.

There is a consequence of this dipole structure; chains of water molecules can associate (Fig. 5.4).

This chain-like structure has an interesting consequence for current flow in ionic systems. To picture this, think about the desktop toy that has several steel balls suspended in a row, touching each other. When one of the balls on the end of the group of suspended balls is drawn back and then released, it strikes the row of touching steel balls and, suddenly, the ball on the opposite end of the row pops off while the row of balls remains stationary. A back and forth then ensues with the end balls swinging out and then returning to strike the row transferring their momentum to the ball on the opposite end. Likewise, a hydrogen ion can associate with one end of the chain of water molecules with the effect that a hydrogen ion pops off the other end of the chain (Fig. 5.5).

When ionic mobilities are determined for various ionic species in aqueous solution, hydrogen and hydroxyl ions have mobilities that are ~5 times faster than those of other species, e.g., sodium ions. Although all charged ions participate in current flow in electrolytic solutions, some traverse the medium much faster than others. But, just because the hydrogen and hydroxyl ions travel the fastest to the electrodes, does not necessarily mean they are the ones involved in electron transfer.

At the cathode, hydrogen ions are the ones accepting electrons and forming molecular hydrogen. If the electric field is maintained long enough, hydrogen gas can be created at the cathodal electrodes (not a desired result in tissue). An anode immersed in interstitial fluid receives electrons primarily from chloride ions and less so from hydroxyl ions. These reactions at the cathode and anode

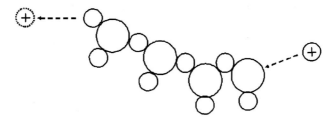

FIGURE 5.5 Depiction of the transfer of a hydrogen ion along a chain of associated water dipoles.

reflect the half-cell potentials of the elements involved in this process. Half-cell potentials are a bit of a complex topic, but essentially it all comes down to which elements are going to be most active in accepting or donating electrons in an electrolytic system. So even though there are many ionic species traveling to the electrodes, only one or two effectively participate in electron transfer. I say effectively because the occasional sodium or potassium ion that accepts an electron at the cathode will immediately form metallic sodium or potassium. These are very reactive entities that will quickly combine with water to regenerate sodium and potassium ions as well as hydrogen or hydronium ions.

This movement of charged ions has another less obvious consequence. As the ions migrate toward electrodes of opposite potential, a shift in ionic concentrations occurs across the solution with charged ions becoming more concentrated in the vicinity of the electrodes over time. This accumulation is facilitated if the electric field is present a significantly greater percentage of time than it is absent because the ions do not have time to return to an equal distribution during the time the field is turned off. If this situation is continued over a period of hours to days, pH changes can occur in the region of the electrodes. This happens as hydrogen ions accumulate near the cathode, and an excess of hydrogen ions constitutes an acid situation. Similarly, hydroxyl ions accumulate near the anode constituting a basic situation. Actually, this process involves the concept of Lewis acids and bases that are defined as electron acceptors and donors and are not restricted to just hydrogen and hydroxyl ions. Therefore, all the positively charged ions accumulating near the cathode function as potential electron acceptors, and all negatively charged ions near the anode function as potential electron donors. This process leading to a local change in pH (the negative logarithm of the hydrogen ion concentration) has been long recognized with electrical field application to tissue, and strategies to avoid damaging the tissue are employed with the electrical systems applying the field. These strategies include adjusting the shape of the waveform to include components of opposite polarity, e.g., tails that cross the baseline in the opposite direction to that of the main pulse or the use of biphasic waveforms such as discussed in the output of Transcutaneous Electrical Nerve Stimulation (TENS) units in the Introduction. Actually, one of the simplest approaches to solving this problem is to turn the system off on a regular basis or to reverse the polarity. When the system is off, any changes in the regional concentration of ions in solution will normalize over time as the ions follow electrochemical gradients to return to an equal distribution. Reversing the polarity of the field actually propels the ions in the opposite direction when the reversal occurs. Recognition that these pH changes that can occur and the adoption of means to mitigate their effects is important in system design for long-term stimulation applications.

Now, with this basic description of the process of electrolytic current flow, we can understand the concept of electrolytic resistance. During this bidirectional movement of charge that constitutes electrolytic current flow the ions interact both repulsively and attractively and those interactions constitute resistance. Fortunately, because of the very large number of ions involved, the resistance of ionic media is described by the following simple relationship in Eq. (5.1):

$$R = \rho(l/A) \tag{5.1}$$

where l is the electrical path length, A is the cross-sectional area of the electrical path, and ρ is a proportionality constant called the resistivity with units of Ohms·cm. This relationship expressed in terms of conductance is demonstrated in Eq. (5.2):

$$G = 1/R = \kappa(A/l) \tag{5.2}$$

where G has the units of Siemens/meter or S/m, sometimes also expressed as mhos/m (mhos for reciprocal Ohms). The reciprocal of resistivity is called conductivity and is represented by the Greek letter kappa ($\kappa = 1/\rho$).

Recall the shape of the curve in Fig. 2.2 that I described as a power function. The setup for those measurements consisted of two ECG electrodes stuck together with polyethylene masks between. Eq. (5.1) shows that because the length of the electric field path did not change, and the resistivity was the same, the determined resistance depended on the cross-sectional area of the holes in the masks. Since area is a power function (πr^2), the form of the curve in Fig. 2.2 becomes obvious. From Eq. (5.1), as the length between electrodes overlying the digital nerve on my thumb increased, unless the cross-sectional area increased to an even greater extent (which it did not), the resistance should increase according to Eq. (5.1), but instead it fell. Since negative resistance being added per unit length of thumb did not make any sense, and several repeats of the measurements provided the same pattern of declining resistance with increasing length, the realization finally dawned that the important factor involved was impedance of which resistance was only a part.

Tissue has traditionally been represented as a parallel RC (p-RC) circuit for which the impedance is again represented by the familiar relationship shown in Eq. (5.3):

$$Z = \left(\frac{1}{R^2} + \frac{1}{X_C^2} \right)^{-0.5} \tag{5.3}$$

where Z is impedance, X_C is the capacitive reactance and is equal to $1/2\pi f\,C$. Clearly, as the capacitance increased with added nerve length in the tissue electrical path, so long as the resistance did not increase proportionally the impedance would fall, and this effect would be more pronounced with increasing frequency of the applied waveform.

Initially a square, constant current waveform output was used for Impedance Neurography studies, mimicking the output of a TENS unit. The maximum value of the output voltage pulse was measured at low frequency (250 Hz) and looked like the depiction in Fig. 5.6.

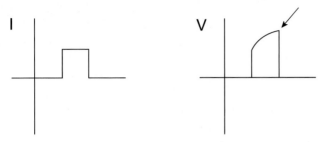

FIGURE 5.6 The current and resultant voltage waveforms from an early Impedance Neurography device. *Arrow indicates* where voltage sampling was performed in the waveform.

Since voltage was measured on the rapidly rising portion of the p-RC charging curve, at pulse termination, significant potential for error introduction was present. The error potential was increased with higher frequency sampling that would shorten the pulse width even more. After much data collection and analysis, it was determined that there was no reason to avoid using a sinusoidal waveform and measuring peak-to-peak voltages to avoid the p-RC charging curve effect. However, a nuance was observed.

The source generator for the Impedance Neurography device employing sinusoidal outputs was constructed to mimic the direct current (DC) offset of the square waveform device and purposely avoided grounding the output between cycles, i.e., though the waveform minimum began as 0 V it was allowed to "float" during the sampling interval. This has been discussed in Chapter 3 and is depicted in Fig. 5.7.

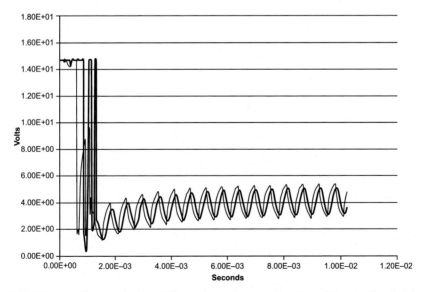

FIGURE 5.7 Square and sinusoidal waveforms from an Impedance Neurography device measured across a tissue path. *(Disclosed in Cory, US Pat. Appl. 20,120,323,134.)*

The floating baseline allowed the output waveform to "climb" the tissue charging curve over the course of the sampling interval, shown in Fig. 5.7, to eventually stabilize. This stabilization effect was more prolonged in the controlled current mode than in the controlled voltage mode. In Fig. 5.7, the heavy black line shows an output sinusoidal waveform, while the lighter black line demonstrates in output square waveform across the tissue path. As expected, both waveforms demonstrated the same effect. If sampling was performed prior to waveform stabilization, the changing baseline was a source of error introduction plus, as discussed in Chapter 3, the peak-to-peak measurements would vary depending on where the sampled pulses were in the stabilizing waveform pattern. Rapid stabilization promoted earlier sampling and facilitated timely image construction.

The settling of the waveforms to a stable value was noted to occur in markedly fewer cycles using a controlled voltage output compared to using a controlled current output as predicted by theory. In recognition of this, controlled voltage outputs became the standard for Impedance Neurography measurements because of the earlier stabilization of the waveform as well as developing more controlled voltage gradients in the tissue path.

Since nerve tissue has higher capacitance than the surrounding tissue, its RC time constant (R × C) is greater and the slope of the RC charging curve will be less than that of the lower capacitance tissue in which the nerve is imbedded (resistivity being similar between the two tissue types) as shown in Fig. 5.8.

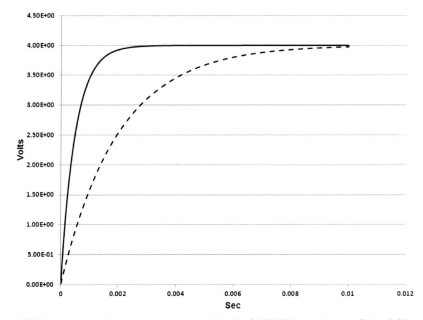

FIGURE 5.8 Two RC charging curves, both with an R of 100,000 Ω, capacitance of the *solid line* equals 5×10^{-9} F, capacitance of the *dashed line* equals 2×10^{-8} F.

The observed settling effect of the applied waveform occurs more quickly with the shorter RC time constant (nonnerve), and sampling of the peak-to-peak voltage from the applied waveform may appear less than that of the longer, nerve-related RC time constant unless care is taken that stabilization has occurred. This is facilitated by using controlled voltage waveforms with fewer cycles required for stabilization. The nature of the electrolytic and electrotonic factors involved in Impedance Neurography was a critical understanding for developing an approach to depth determination capability with the technology.

THE GOAL OF DEPTH DETERMINATION

When we first realized that it was possible to determine the course of nerves at the skin surface, the question arose regarding being able to accurately determine nerve depth and create three-dimensional (3-D) maps in addition to the two-dimensional (2-D) maps. On initially approaching this problem, it was proposed to us that the same mathematical technique used in Impedance Tomography be adapted to Impedance Neurography: finite element analysis or finite difference analysis. This approach is based on some assumptions regarding how current and voltage were being distributed in tissue. These assumptions are not commonly discussed in works relative to Impedance Tomography, but an understanding of them is critical for recognition of the differences between the two technologies. The primary, tacit Impedance Tomography assumption is that tissue is best represented, from an electrical standpoint, as a homogeneous, bulk conductor. An attractive aspect to this approach is that one is working with a "black box" the internal structure of which is not necessarily important. Rather, the surface voltage distributions can be mathematically manipulated to provide a model of the interior of the box. We proceeded down this track for a while, but with reflection on the anisotropicity discussed in Chapter 4, it became apparent the finite element or finite difference analyses would not be productive. Since Impedance Neurography makes determinations of the impedance of the electrical path, knowing the structures contained in that path and how they relate one to another is a key part of modeling the system.

At one point in the Impedance Neurography journey, I was invited to demonstrate the technology to a company based in Jena, Germany. It was an interesting visit for a number of reasons. One was that I stayed at an old inn that turned out to be a location where Martin Luther held meetings for various reasons including to debate Andreas Karlstadt in 1524. The room in which those meetings took place was preserved with the same furniture and accouterments. Wandering the environs of the Hotel Schwarzen Bar, including the University of Jena, I bumped into Hegel, or at least a bust of Hegel. He taught Philosophy at the University, first as an unsalaried lecturer and then as an Extraordinary Professor, also unsalaried. Some things never seem to change very much. A small cemetery a few blocks away held the discovery of a beautiful, pyramidal monument to Carl Zeiss, and I learned Napoleon fought a battle with the Prussians on the plateau not far from Jena city center. I had no idea of the rich history I would encounter when I started off for Jena.

At the meeting, the demonstration did not go particularly well since I was still unaware that sampling at 2 kHz was nonoptimal. At one point, I showed a meeting participant a series of calculations in which I was engaged to try and determine a correlation with depth. He remarked that it looked to him like I was guessing, though the equations were derived from multiple circuit models, all of which could have played a role in how the field distributed in tissue. Since things were not going too well anyway, I did not feel there was anything to be gained by trying to explain further, especially since he was not mathematically inclined. But these kinds of determinations are important for depth determination and 3-D image reconstruction. It turned out that the approach of just selecting multiple circuit configurations and testing the mathematical descriptions of those circuits against known depth information was not fruitful.

What we really needed was a way of defining the relationship of the skin surface impedance determinations and the underlying anisotropic electrical path: the nerves. Forming a mental image of the structures involved in the circuit was an important step forward enabling the determination that three portions of nerve were involved in the electrical path: (1) the small nerves running from beneath the sampling electrodes on the skin down to (2) the larger and deeper main nerve, and then (3) the small nerves running up to the skin under the return electrode. This was discussed in Chapter 3 in the determination of the equivalent electrical circuit. This visualization process has been used many times in the development of Impedance Neurography, and it helps avoid an important stumbling block. All too frequently, when attempting to define how a system is working, previously developed mathematical models are adapted to the new situation. Though often useful, applying an analogy between a standard model and new situation may be a flawed approach. This was certainly the case for Impedance Neurography where the standard model depended on the tacit assumption of bulk conductor characteristics of tissue.

What became apparent from the equivalent circuit model was that the current distribution in living tissue was not diffuse, but rather, it was discrete and followed specific anisotropic paths that had been detected empirically from the very first observations. We also noted from the very beginning that a gradient of impedance values occurred on either side of the lowest impedance line demarcating on the skin surface the course of the underlying nerve. The existence of the impedance gradient strongly suggested a relationship between the skin surface measurement site and the distance to the underlying nerve. The question was what comprised that relationship.

DIMENSION FACTORS IN AXON IMPEDANCE

The impedance reduction effect of increased length of nerve between sampling electrodes led to a confusing observation. As sampling was performed laterally from the course of a nerve, the impedance increased, but if small nerves were providing an electrically anisotropic path down to the underlying large nerve, the current should reach the larger, underlying trunk by traveling along the smaller fibers from the skin surface. If this was correct, then as sampling moved laterally, more nerve path length was added and one might anticipate that the overall impedance would fall, but this was not observed. This situation was shown in Fig. 2.5 where an additional finding was obtained from comparison studies of MRI and Impedance Neurography images. The depth of imaged nerves was determined from MRI studies and shown to be unrelated to the direct path between the nerves and the skin surface measurement sites. Interestingly, the depth was found to correlate with the sum of the two, nonhypotenuse sides of a right triangle of which the direct path constituted the hypotenuse, and the base was represented by the lateral distance from the minimum impedance site on the skin surface to the sampling electrode. This finding corroborated the early, and consistent, finding that nerves were always found on a normal to the site of lowest impedance on the complex plane of the skin surface, i.e., a right-angle relationship. The projection on a normal was confirmed on several other MRI comparison studies as well. The reason for confusion about the right-angle relationship is shown diagrammatically in Fig. 5.9.

An anatomic substrate to explain the right-angle relationship was not readily apparent from the literature. Though small nerve fibers run at right angles to the skin surface from immediately below the dermis in the last millimeter or so, there is no known right-angle anisotropic structure to explain the larger dimension down to the underlying nerve. So, what was happening?

Solving this question turns out to involve thinking about junior high geometry. In Fig. 5.10, the nonhypotenuse sides of a right triangle, a + b equal 5(c + d). But as the steps become ever smaller, it's tempting to think that the sum of the lengths of the nonhypotenuse sides will eventually become the same as the length of the hypotenuse itself. This never happens no matter how small

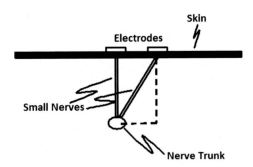

FIGURE 5.9   The presumed courses of two, small nerve fibers running from an underlying nerve to the skin. The *dashed lines* represent the right-angle relationship that was found to correlate with depth.

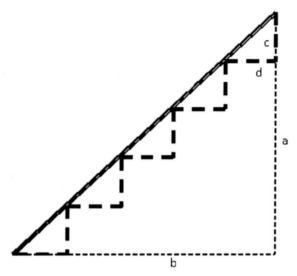

FIGURE 5.10   Representation of the varieties of the right-angle path.

the steps become, though the multiplier changes as the steps become smaller and the stair step approximates the course of the hypotenuse. Importantly, no matter how small the stair steps become, their sum will never be the same as the length of the hypotenuse, but always equal to the sum $a + b$. This appears to replicate how a nerve fiber courses between cells, or tissue planes, i.e., the nerve fiber stair steps around the cells or tissue planes rather than poking through them as they would have to if following the hypotenuse of the right triangle.

This also fits with how nerves develop embryologically in that the tissues form first, then nerves migrate along nerve growth factor gradients to reach their target tissues for innervation. For developing nerves to follow those chemical traces that lead them to their targets, they have to track around cells in the interstial spaces where the chemicals reside.

Although the right-angle relationship can be explained by the above, this does not account for the increased impedance with lateral positioning of the skin surface measurement site from the normal between the nerve and complex plane of the skin surface. Electrotonics provides the answer to this question.

Electrically, using Impedance Neurography equipment with a charged-DC (c-DC) offset, nerves constitute a p-RLC (parallel Resistance Inductance Capacitance) circuit discussed in Chapter 3. The impedance of a nerve fiber can be described mathematically as Eq. (5.4):

$$Z = \left(1/R^2 + 1/X_C^2 + 1/X_L^2\right)^{-0.5} \tag{5.4}$$

where X_L is the inductive reactance ($X_L = 2\pi Lf$). If R is very large compared to C and L, it becomes the dominant component of the impedance, Z. Why is this so? The answer has to do with the nature of parallel resistance, inductance, and capacitance.

The total resistance of parallel resistances is found as in Eq. (5.5):

$$1/R_T = (1/R_1 + 1/R_2 + \ldots + 1/R_n) \tag{5.5}$$

Using just two resistances, $R_1 = 1000\ \Omega$ and $R_2 = 1{,}000{,}000\ \Omega$, the parallel resistance will be calculated in Eq. (5.6):

$$R_T = (0.001 + 0.000001)^{-1} = 999.001\ \Omega \tag{5.6}$$

What is clear is that the total resistance will approximate the lowest resistance in the group of parallel resistances, and if all the resistances are the same, the total resistance will equal the same quantity as each individual resistance. This makes sense because if a circuit has two, parallel resistors of equal magnitude, $\frac{1}{2}$ the applied current will course through each resistor, which is equivalent to all the current traveling through one of the resistors, i.e., the total resistance is the same.

Similarly, the total inductance of a group of parallel inductances is calculated the same way as the total resistance, e.g., Eq. (5.7):

$$1/L_T = (1/L_1 + 1/L_2 + \ldots + 1/L_n) \tag{5.7}$$

Thus, if a group of identical, small axons are arranged in parallel, and since the axons are identical in my *Gedankenexperiment*, the total resistance and total inductance will be the same as the resistance and inductance of any, individual axon.

It is the total capacitance that really affects the impedance of such an arrangement of axons. The reason is that the total parallel capacitance is additive shown in Eq. (5.8):

$$C_T = C_1 + C_2 + \ldots + C_n \tag{5.8}$$

The additive nature of capacitance makes sense since as more capacitive surface is added to the system, the total capacitance is expected to rise; there is more capacitive surface on which to store charge. For a bundle of x axons that are identical, the total capacitance is equal to the product of the number of axons and each individual capacitance. We know that if the total resistance and inductance do not change as more ideal axons are bundled together, but the capacitance rises arithmetically, the impedance will fall due to the reciprocal nature of the effect of capacitance on impedance.

For small nerve fibers (axons) that are typically less than $10\ \mu$ in diameter, the internal resistance represented by the neuronal cytoplasm is much greater than the membrane-associated capacitive and inductive reactances, though this is length related, and at sufficient length the more typical exponential decay of impedance versus length is seen as shown in Fig. 5.11. As an impedance source, small, relatively short fibers will act similarly to a linear resistive material where incremental length increases result in incremental resistance increases.

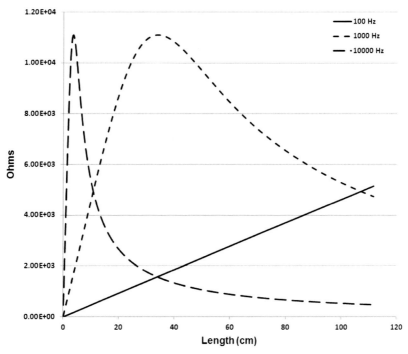

FIGURE 5.11   Three calculated impedance versus length relationships for an idealized, 6 μ axon with an intracellular resistivity of $100 \, \Omega \cdot$ cm.

Fig. 5.11 also shows a frequency-dependent, linear initial portion of the impedance versus length relationship for a single, 6 μ axon represented as a p-RLC circuit. The frequency dependence of this effect may be used to advantage as seen in a plot of Impedance Neurography impedance data versus electrode position, Fig. 5.12 where eight Impedance Neurography electrodes in a row crossed, at right angles to, the course of the saphenous nerve of the calf. Similar magnitude impedance values are seen from neighboring electrodes on each side of the minimum impedance electrode.

In Fig. 5.12, the farther lateral an individual electrode resided from the position of the nerve, the higher was the calculated impedance. Electrode #6 lay approximately on a normal to the position of the saphenous nerve, and an impedance gradient was seen to either side of that electrode. This effect is most pronounced at lower frequencies and is essentially lost at frequencies above 6000 Hz.

Considering the dimensions of the electrodes in the array allows for additional definition of the point on the skin surface located precisely on the normal connecting the nerve to the complex curve of the skin surface. The electrodes in the array had a 3 mm diameter and a 5 mm center-to-center

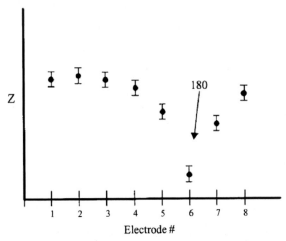

FIGURE 5.12   Impedance versus electrode position obtained over the saphenous nerve. *(Disclosed in Cory, US Pat. Appl. US20120323134.)*

spacing. The edge-to-edge spacing between the electrodes was 2 mm, and a normal connecting the skin surface to the nerve could easily lie between two electrodes in the array. This is suggested in Fig. 5.12 noting that the impedance values for electrodes 7 and 8 are slightly lower than for electrodes 5 and 4, respectively. This observation implied that the actual position of the underlying nerve was displaced to a point between electrodes 6 and 7.

Referring Figs. 5.7 and 5.10, the small fiber resistance contribution to the overall impedance measurement allows for the differentiation of electrodes positioned more laterally to the course of the underlying nerve.

However, as more and more axons are bundled together, the total impedance will fall due to the parallel nature of the nerve anatomy causing the reactive components of the impedance to dominate, explaining the observation of falling impedance with electrode separation distance along the course of a peripheral nerve.

Fig. 5.11 provided data related to a single, idealized axon. Figs. 5.13 and 5.14 provide data from 10 to 10,000 of those idealized axons. Note the different length dimensions between Figs. 5.11, 5.13, and 5.14.

Figs. 5.13 and 5.14 demonstrate the effect of the highly parallel nature of the axons comprising nerves on the measured impedance. With more axons running in parallel, the resistive portion of the impedance plays less of a role than the reactive portion. Additionally, lower frequency waveforms used for sampling will accentuate the small fiber, linear resistive component, i.e., at lower frequencies, greater lengths of small fibers will display linear resistance. The effect will be lost as frequency increases.

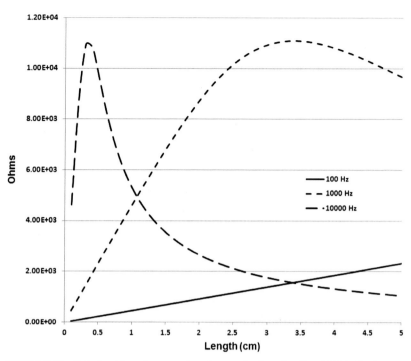

FIGURE 5.13   Calculated impedances at three frequencies for a bundle of 10 idealized axons.

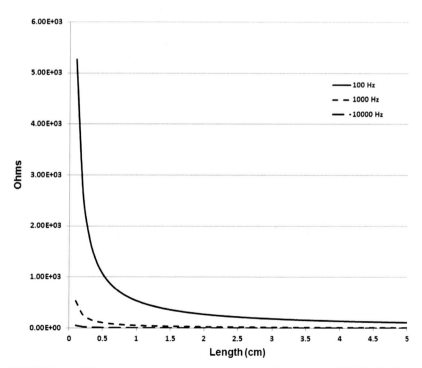

FIGURE 5.14   Calculated impedances at three frequencies for a bundle of 10,000 idealized axons.

In Figs. 5.9 and 5.12, it is an important consideration that the electrical path for an array of sampling electrodes coupled to a single return electrode is the same for all the electrodes in the array except for the final segment from the larger, underlying nerve to the individual electrodes. That final path varies depending on the position of the individual electrodes in the array and the calculated impedances reflect the contribution of the small fibers coursing to the skin surface in addition to the common pathway of the larger nerve. As the path length along the small fibers increases, the impedance rises concordantly.

Recognition of the linear resistive component tracking from the underlying nerve to the sampling electrode that correlated with sum of the nonhypotenuse sides of a right triangle was tantalizing as it seemed to offer a very simple means of depth determination. This was found to be true in some experiments, but not in others. Ambiguity was rearing its head once again.

Though tempting to disregard the times when very accurate depth determination was possible using the geometric relationships described above as merely coincidence, I find that explanation more useful as a plot device in fiction. The significant observation is not the times depth determination did not work, but those when it did. What was going on?

I do not have a complete explanation for this question, but there are intriguing observations that may point the direction for future study having to do with the most variable part of the final path from the underlying nerve to the surface electrode: the skin. This organ constitutes the barrier between us and the outside world. It is amazingly tough and resists a host of insults including chemical, infectious, electrical, and environmental (e.g., drying). It also is a very tempting target for therapeutic interventions and for millennia, unguents, potions, emollients, and various other forms of goop have been touted beneficial when smeared on the skin. Some, in fact, are capable of penetrating the skin usually after prolonged contact. A substantial amount of effort has been invested in finding ways of breaching the skin defenses for therapeutic applications (or toxic ones such as assassinations), and patch systems for drug delivery are now common. Sometimes drug delivery is accomplished with the assistance of electrical fields, a technique called iontophoresis. In this application, charged molecules can be "driven" through the skin following an electric field of appropriate polarity. And it is clear from all the foregoing material that electric currents can get through the skin without damaging it if the current levels are kept below a safe minimum value. But there is an interesting question that comes up when thinking about current getting through the skin; how does it do it?

As part of answering this question, I came across the very interesting work of Dr. Grimnes[1] who constructed a novel experiment to look at this question. By placing very thin metal electrodes on dry skin (thin enough to see through), then passing a current through the skin to the electrode, he noticed that small dots formed on the electrode indicating regions of high current density. Those small dots were found to highly correspond to the location of sweat gland orifices. Now this makes a lot of sense given that dry skin has a very high impedance and a fluid-filled sweat gland offers a low impedance path through the high impedance stratum corneum. Furthermore, a parasympathetic nerve resides at the base of each sweat gland suppling innervation to the gland. We know about the ability of nerves to provide facilitated conduction through tissue, so when coupled with the sweat gland, the combination seems a good candidate for the electrical path through the skin. However, we cannot depend on biological structures always comporting with our view of how the world works, and sweat glands are no different in that regard. The orifice of a sweat gland can close, so at any one time, the population of sweat glands in a given area of skin may be in various states of openness. This implies that the glands may also be in various states of ability to conduct electrical current and that situation contributes to the variability of impedance measured at the skin surface. So, what can be done to improve the situation?

We know from experiments using prolonged electrode contact with the skin that impedance continually declines. I got tired of the experiment after 6 h, noting that the impedance was still declining even after that long time period. This observation is a reflection of skin hydration that was mentioned earlier. That is a different process from the sweat gland contribution to skin impedance and if some way could be found to rapidly affect sweat gland function and stabilize the glands in the open position, it could be very advantageous for Impedance Neurography. The concern is that being informed by the three-part series circuit as described in Chapter 3, and the necessity of intact skin for effective Impedance Neurography measurements, skin impedance reduction by affecting sweat gland function might be counterproductive. All of this brings us to a topic not often discussed in polite society: body odor.

As a society, we spend significant amounts of money trying to prevent body odor and, probably, bathe more frequently than is good for us. Besides soap and water, one of the strategies we use for smelling good is antiperspirant application. For a long time, it has been recognized that bacterial decomposition of protein components of sweat leads to unpleasant odors, so we try to reduce sweating with these products, and they work. The question is, *how* do they work?

Antiperspirants have been around since the 1800s. The first agent recognized to be effective in reducing sweating was aluminum chloride, and since then zirconium chloride has become more the industry standard. So, what is it about aluminum and zirconium that result in decreased sweating when these agents are applied topically? The short answer is no one knows for certain. There is, however, a longer answer.

Aluminum and zirconium both form trivalent cations in solution. That means that one aluminum atom can bind with three monovalent anions, such as chloride, giving the chemical formula of $AlCl_3$ for aluminum chloride. That is not its claim to fame in antiperspirant lore. Since aluminum can bind to three other negatively charged entities, it is a good cross-linking agent. This means that rather than just making linear chains of molecules of the form:

$$x - x - x - x - x - M - x - x - x - x - x$$

aluminum can make more complex combinations, e.g., Fig. 5.15.

This ability to cross-link other molecules together has led some to think that trivalent cations exert their antiperspirant effect by creating physical plugs of cross-linked proteinaceous debris in the sweat gland pores. There are even electron micrographs reputed to show these plugged-up pores. The only problem is that it's impossible to show an electron micrograph of a pore prior to exposure to aluminum or zirconium, and then after exposure due to the way tissue has to be handled for electron microscopy. Skin stripping experiments using tape to peel off the stratum corneum following antiperspirant application and even Raman spectroscopy have been used to try and define whether these plugs exist. At present, no one really knows whether the debris plug hypothesis is, in fact, correct. As you might have guessed, there is a competing point of view (you never thought antiperspirants could be so interesting, did you?).

Some people perspire excessively and have a syndrome called hyperhidrosis that is socially difficult for them, especially in situations where they are stressed and do not have a suit coat inside of which to hide. There are various treatments for hyperhidrosis, including Botulinum toxin injection and surgery in extreme cases. Most situations can be handled more simply with agents such as Drysol.

Drysol is a 20% solution of aluminum chloride hexahydrate in absolute alcohol that is applied to the skin via an applicator very similar to the familiar

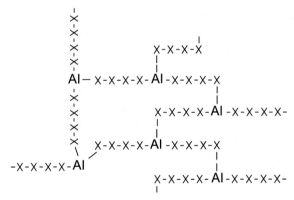

FIGURE 5.15   Aluminum atoms cross-linked to other, organic molecules.

antiperspirant applicators. For some individuals suffering excessive sweating, when applied at regular intervals, it can result in a marked decrease in sweating for the long term, e.g., months to years. For reasons that are not easy to explain, I thought it might be a good idea to apply this agent to skin and then check impedance measurements; I still do not know why that occurred to me except that it was the only topical agent I knew of that affects sweat gland function. Fig. 5.16 shows the results.

The experimental setup for collecting the data in Fig. 5.16 began with selecting a patch of skin on the volar forearm where two electrode arrays could be positioned side-by-side, i.e., the skin sites were as identical as possible. Prior to applying the arrays, the area over which one array was to be applied was gently wiped with Drysol and the alcohol allowed to evaporate. While wiping, it was important not to be heavy-handed since removal of the stratum corneum would skew the results. Controlling for this was accomplished by gently wiping the other electrode application site with absolute alcohol that did not contain aluminum chloride. Then, the two electrode arrays were applied, simultaneously, and sampling begun immediately. More frequent sampling was performed during the first 10 min of application time, then at 5 min intervals. What was obvious from this experiment, as well as others, was that a rapid and marked decrease in skin surface impedance determinations occurred in the Drysol prepped skin versus the control prepped with alcohol alone.

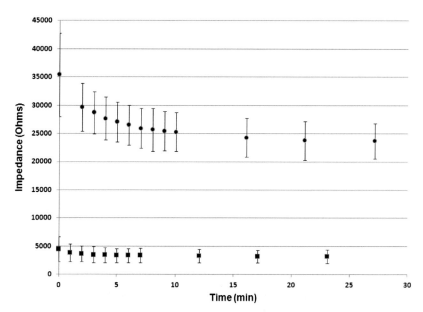

FIGURE 5.16   Time course experiment looking at impedance determinations of alcohol prepped skin (*closed circles*) and Drysol prepped skin (*closed squares*). (*Data courtesy of Nervonix, Inc. Graphic constructed by Philip C. Cory, M.D.*)

What do the data in Fig. 5.16 mean? If the sweat glands were being physically plugged by aluminum cross-linked debris, the impedance would not be expected to fall but would likely increase due to loss of a fluid-filled path for current flow down the sweat gland pore. Instead, there is a reproducible and marked fall in total system impedance to a minimum value with much smaller standard deviations than in the control situation. It appears that trivalent cations, rather than resulting in physical plugs in sweat gland pores, have a pharmacologic or even toxic effect on sweat gland fluid production, while at the same time leaving the population of pores widely patent, maximizing the area for current flow through the skin.

Notably, application of alcohol without aluminum chloride did not cause the same effect. This stabilization of skin surface impedance conditions may make the assessment of the right-angle relationship of impedance to nerve depth much easier and more reproducible. Those experiments remain to be performed.

There was another interesting observation from the Drysol experiments. The electrode array used for those measurements had 60 electrodes with a skin attachment system that contained a hole for each, individual electrode. This skin attachment system was constructed from closed cell, acrylic foam that was 2 mm thick. The holes were 3 mm in diameter and filled with an aqueous conductive gel that provided the interface between the electrode surface and the skin surface. Usually, after a measurement session, when the electrode array and skin attachment system were removed, 60 little dots of gel were left on the skin surface. This was not seen when the system was removed from the Drysol prepped skin. Instead, there were 60, raised skin bumps corresponding to the holes in the skin attachment system and most of the gel was missing from the holes, but the missing gel was nowhere to be seen on the skin surface. The little skin bumps had actually been drawn up into the holes in the skin attachment system. The Drysol prepped skin may have imbibed the water from the gel, whereas the neighboring patch of skin that had not been similarly prepped was festooned with the little gel dots commonly seen on electrode array removal. The same skin bumps were seen when an unfilled, control skin attachment system was placed on the skin surface. My conclusion was that not only did the sweat glands not produce fluid but if they provided a path for fluid imbibition into the skin, that fluid evacuation from the skin attachment system was responsible for the appearance of the skin bumps. It was as though the loss of fluid made the initially full skin attachment system on the prepped skin behave like the empty one over time.

Considering how the fluid moved into the skin is interesting because of the relationship of the molecular concentrations involved. Water typically moves from underneath the skin to the skin surface where it evaporates as a means of eliminating heat associated with metabolism. We are familiar with this process when we perspire, but it occurs all the time since water is concentrated beneath the skin in the interstitial fluid and moves toward the skin surface where the water concentration in the atmosphere is much lower. This is an example of water movement along a concentration gradient. This movement is impeded when a Band-Aid, or similar device, is placed on the skin, evidenced by the swelling of the skin and lightening of its color that develops the longer the Band-Aid is in place. These changes are a result of the prevention of water evaporation and its accumulation in the stratum corneum.

The wet gel used for Impedance Neurography experiments was Redux Paste, which is quite hypertonic, having a specific gravity of 1.13 compared to that of serum or blood, which is around 1.03−1.06. The hypertonicity of the Redux Paste should have "drawn" water out of the skin to dilute the gel, but the opposite occurred. The reason for this is that the gel contains a high concentration of sodium chloride to make it very conductive. In fact, we chose Redux Paste based on conductivity measurements that were performed by our collaborating engineers on several commercial medical gels and it was found to be the most conductive. The high sodium chloride concentration was reflected in how much the gel stung if it happened to get on a scratch in the skin. Since the concentration of the sodium chloride was much greater on the gel side of the skin, the sodium and chloride ions would tend to migrate along their concentration gradient though the stratum corneum toward the dermis. Normally, this was blocked by the continuity of the skin barrier, but by keeping the sweat glands in the open position, due to the aluminum ion effects breaching skin continuity, the migratory ions had a path to travel from exterior to interior. It appears that water molecules passively followed the ions as they moved into the skin, draining gel out of the holes in the skin attachment system. If this mechanism is a correct explanation for the observed effects, the use of trivalent cations to promote absorption of therapeutic molecules through the skin may be possible and merits additional investigation.

IMPEDANCE VERSUS ELECTRODE SPACING

This recognition of the impedance contribution of the small fibers directly beneath the sampling electrodes offers an explanation for another finding that has been a source of confusion in Impedance Neurography; the impedance versus electrode spacing observations shown in Fig. 5.17.

Fig. 5.17 is a graph of impedance versus electrode separation distance over the course of the saphenous nerve. There are a couple of things to be seen in these data.

The sites sampled at 1.5 cm lateral to the saphenous nerve consistently demonstrate impedance values that are higher than those measured directly over the nerve. This is the right-angle relationship, recognized from the earliest investigations.

Additionally, there is a marked relationship between sampling and return electrode separation distance and impedance. At electrode separations less than 15 cm, the impedance values are much higher than at electrode spacing

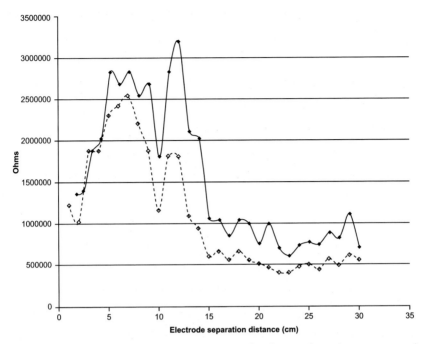

FIGURE 5.17 Data from two electrodes positioned directly over the saphenous nerve and laterally at 1.5 cm to the saphenous nerve. Impedances were calculated as the electrode separation distance was increased from 1 to 30 cm in 1 cm increments. *(Disclosed in Cory, US Pat. Appl. US20110082383.)*

greater than 15 cm. Then, there is the region with less than 5 cm spacing where impedance values are lower. Referring to Fig. 5.18 helps explain these observations.

As the two surface electrodes in Fig. 5.18 are spaced more closely, the length of larger underlying nerve that lies in the electrical path decreases and the percentage contribution of the small fibers to the measured impedance increases, also changing the contributions of resistance and reactance to the calculated impedance values.

Fig. 5.17 demonstrates a rather abrupt transition from the resistance-dominant portion of the impedance versus electrode separation distance relationship to the reactance-dominant portion. This occurred at around 12–15 cm for the saphenous nerve in Fig. 5.17. The lower impedances observed at separations of less than 5 cm may reflect that the anisotropicity of the nerves does not supply an advantageous electrical path at that distance and that the bulk conductor characteristics of the tissue become more critical. If that is the case, it explains the apparent loss of nerve discrimination at the short electrode separation distances.

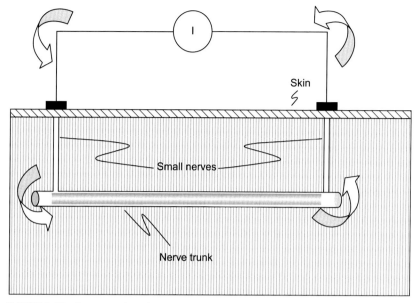

FIGURE 5.18   Representation of nerve electrical anisotropic pathway for current flow in tissue. *Arrows* show current flow preferentially along nerves.

THE LAST ISSUE FOR EFFECTIVE THREE-DIMENSIONAL IMAGING

The ability to use skin surface impedance determinations resulting from the application of a time-variant electric field superimposed on a c-DC offset for the construction of 2-D maps of nerve position has been well-documented in preceding chapters. Also, there is the distinct possibility of using these same impedance values for determining the depth of nerves underlying the skin surface. One thing remains to be solved to effectively develop useful 3-D images of nerves, which is a means to determine the size of the nerve. Knowing both the dimensions of the underlying nerve as well as its depth is critical to constructing an image of the nerve in a block of tissue. The remaining issue is whether there is enough information available from the skin surface impedance values to answer the dimension question.

Recall Figs. 5.13 and 5.14 showing impedances for bundles of nerves containing 10 axons and 10,000 axons at various frequencies. The shapes of those curves were quite different because of the number of parallel impedances associated with the axon numbers. This information may well offer a method to determine nerve dimensions.

In Fig. 5.19, the impedances of bundles of axons from 10 to 5,000 axons were calculated at three different frequencies: 50, 75, and 100 Hz.

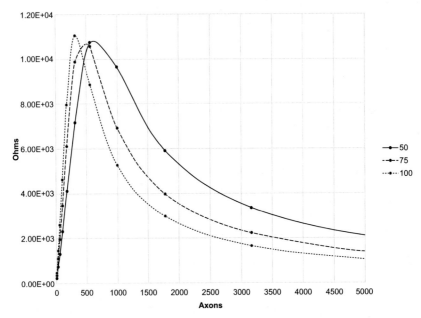

FIGURE 5.19 Impedance calculations for axon bundles from 10 to 5,000 axons at three different frequencies of 50, 75, and 100 Hz.

As expected from the above discussions, the impedance profiles are different for the three different frequencies. This implies that frequency scanning may provide sufficient information to make determinations about the numbers of axons participating in the circuit and the dimensions of the nerve containing those axons. This, too, remains an area for future researches and is significant in regard to the final section looking at fusion technologies.

FUSION TECHNOLOGIES

The ability to determine the depth of peripheral nerve fibers, using the right-triangle relationships described above, enables construction of 3-D blocks showing the position of embedded nerves. This capability is useful on its own but provides a means to enhance the information available from other technologies.

Consider imaging that employs X-rays. This includes plain radiographs, e.g., chest X-rays, mammograms, and computerized systems, such as CT scans. These technologies have no ability to visualize nerve tissue, but what if an additional, straightforward imaging system that was specific for nerve could be combined with these other technologies? Nerve tissue could be displayed, in real time, with X-ray-based systems such as fluoroscopy. Such a fusion of technologies would allow targeting of nerve tissue or avoidance of nerve tissue during procedures. Likewise, Impedance Neurography could be combined with MRI or ultrasound to indicate which structures were nerve and which were vessels or other linear structures, recalling the MRI image of the sciatic nerve (Fig. 2.5). And for those who have struggled with determining the position of nerves from the gray-scale raster of ultrasound images, having a dot placed in the appropriate position would be very helpful. These fusion techniques could be very valuable adjuncts to surgical procedures in the operating room, or during planning of therapies that would profit from knowledge of nerve location.

REFERENCE

1. Grimnes S. Pathways of ionic flow through human skin in vivo. *Acta Derm Venereol (Stokh)* 1984;**64**:93−8.

Appendix

Epilogue

In this book, I have discussed the early clinical and experimental findings that led to the successful construction of an Impedance Neurography device. This equipment provided images of nerves for which there was no explanation in the extant literature and about which I was clueless as to the underlying electro-physiologic mechanism. It was this inability to explain the findings with reasonable hypotheses that made development of a commercially available device difficult. Over 25 years, enough data have been accumulated and been analyzed to enable the construction of a theory of Impedance Neurography that appears to explain the findings that were so confusing initially. As happens in many research projects, coming to an understanding of the underlying electroneurophysiology of Impedance Neurography also provides insights into other therapeutic modalities, in this case nerve stimulation, neuromodulation, and possibly even true electro-anesthesia from the observation of critical factors for inducing electrical resonance in neuronal cell membranes. Seeing how these early findings are used in the future for such applications will be most interesting. Some of what I have written will be found inaccurate or frankly wrong, and that, too, will be interesting. This is the great appeal of scientific discovery; that no one ever has the whole story and there are vistas yet to be seen.

Scientific discovery comes in two types. One is the investigation of territory glimpsed, but unexplored. This involves poking into all the nooks and crannies to see what may be found therein, and sometimes it's treasure. The other type is *finding* that new territory, which is a thrill to which all of us can relate. It's akin to pushing through the clothes to the back of the wardrobe and finding Narnia, or to be Robert Conway climbing across the wind-swept glaciers and snowfields to discover Shangri-la. This is exciting stuff and all it requires is the ability to see the world as exciting and mysterious, and to be able to ask the question, "How is *that* happening?" Too often the innate curiosity with which all of us are endowed gets unwittingly suppressed by the drudgery of daily life, or willfully by succumbing to the allure of our technological toys.

And, most importantly, if you have an observation that is real, but unexplained, do not give up just because one or more people tell you it does not work. I heard that a number of times from well-trained and knowledgeable people who assumed their notion of how the world works was complete. Turns out there were some things they did not know. Be prepared that it might take a decade or two to figure it out... and do not quit your day job in the meantime.

Index

Note: "Page numbers followed by "f" indicate figures, "t" indicate tables."

Printed in the United States
By Bookmasters